Lecture Notes in Physics

Edited by H. Araki, Kyoto, J. Ehlers, München, K. Hepp, Zürich
R. Kippenhahn, München, H. A. Weidenmüller, Heidelberg
J. Wess, Karlsruhe and J. Zittartz, Köln
Managing Editor: W. Beiglböck

308

Henri Bacry

W0192991

Localizability and Space in Quantum Physics

Springer-Verlag
Berlin Heidelberg GmbH

Author

Henri Bacry
Centre de Physique Théorique, C.N.R.S., Case 907
F-13288 Marseille, Cedex 9, France

ISBN 978-3-662-13665-2   ISBN 978-3-540-45908-8 (eBook)
DOI 10.1007/978-3-540-45908-8

This work is subject to copyright. All rights are reserved, whether the whole or part of the material
is concerned, specifically the rights of translation, reprinting, re-use of illustrations, recitation,
broadcasting, reproduction on microfilms or in other ways, and storage in data banks. Duplication
of this publication or parts thereof is only permitted under the provisions of the German Copyright
Law of September 9, 1965, in its version of June 24, 1985, and a copyright fee must always be
paid. Violations fall under the prosecution act of the German Copyright Law.

© Springer-Verlag Berlin Heidelberg 1988
Originally published by Springer-Verlag Berlin Heidelberg New York in 1988
Softcover reprint of the hardcover 1st edition 1988

2158/3140-543210

To

Yael
Hélène
Jacques-Daniel
Emmanuel

SUMMARY.

ACKNOWLEDGMENTS

I am grateful to Professor Louis Michel for his hospitality at the *Institut des Hautes Etudes Scientifiques* at Bures-sur-Yvette, where this work started. Part of the actual writing was done at the *International Centre for Theoretical Physics* in Trieste, for which I am indebted to Professor Abdus Salam. I must also thank Professors L.C. Biedenharn (Durham), L.N. Chang (Blacksburg), N.P. Chang (New York), L.P. Horwitz and Y. Ne'eman (Tel-Aviv), J. Patera (Montréal) and J. Zak (Haifa), who gave me the opportunity to give talks on the subject and to have stimulating discussions with various people.

Many thanks are due to those who followed or helped my efforts in writing the present work, namely A.J. Bracken, D. Kastler, L. Michel, J. Zak and especially to L.C. Biedenharn and A. Grossmann, who also read and criticized the preliminary text. I owe a particular debt of gratitude to A. Connes. This work would never have been written without his enthusiasm in the enlightening discussions we had.

"Ce serait pourtant de l'optimisme exagéré de croire que le moment analytique soit disparu et définitivement remplacé par la synthèse. Il est encore loin de sa disparition et ne fait qu'empêcher le développement de la racine synthétique".

Vassily Kandinsky,
XXe siècle, 3 (1938).

INTRODUCTION.

The aim of the present lectures is to show the reader why the Minkowski space-time is not satisfactory for particle theory. Although the text corresponds to a critical work on quantum theory, it is written at the graduate level in such a way that it can be understood even by a physicist not working in particle physics. My hope is to convince the reader, especially the theoretician, to examine with critical eyes the present situation of particle theory and quantum electrodynamics. Perhaps in going along the lines which are given here, he will be able to propose new directions towards a better theory.

The framework is the one of standard quantum theory and special relativity. It is shown that, if we want to consider seriously the concepts which are the *necessary* ingredients of these theories, we are obliged to give up some superfluous ideas or principles which are usually accepted. The fact that these ideas are incorporated in modern textbooks is only of historical interest. The reader will see why we have to give up:

i) Minkowski space-time (but not special relativity, the energy-momentum space, the Poincaré invariance),

ii) the complementarity principle,

iii) the wave-corpuscle duality,

iv) the quantization procedure as a universal way of constructing the quantum theory.

The reader will not be astonished that the concept which constitutes the core of the present lectures is the one of the *photon* . After all, electrodynamics is the discipline which governs all our observations. That explains why I felt necessary to devote a whole chapter to a historical sketch on the nature of light. The photon and the space are certainly among the most difficult concepts in physics and it was necessary to quote very often our great masters about them. I made my best to provide the readers with the original quotations. English translations of these quotations are provided at the end of the book.

Almost all the great physicists who initiated quantum physics during the first thirty years of our century have disappeared: Planck, Einstein, Bohr, Born, Schrödinger, de Broglie, Dirac, Heisenberg, Jordan, Pauli, Yukawa, etc. Almost all of them left our world deeply unsatisfied by the state of quantum theory[1]. If the reader wants to know

---

[1] This is well known in the case of Einstein, de Broglie and Schrödinger but less known in the case of Dirac:*"There are great difficulties...in connection with the present quantum mechanics. It is the best that one can do up till now. But, one should not suppose that it will survive indefinitely into the future."* He also says: *"Our present quantum theory is very good, provided we do not try to push it too far... I feel that the foundations of quantum mechanics have not yet been correctly established."*(confer [Dir] in the

more about their feelings, he could read their correspondence or he could ask one of the great alive founders, Wigner, or one of the younger masters of physics like Schwinger.

Despite of this lack of satisfaction and despite of their fights (there were really strong fights!), they were very creative and successful as everybody knows. It seems to me that younger physicists must be taught about those fights and the deep critical arguments against quantum theory, not only because they have to be aware of the difficulties of the theory they are familiar with, but also to learn how they could make improvements by themselves in following the steps of the great pioneers. One of the best last examples of such a behaviour is the one of Bell who discovered his inequalities, a source of new beautiful experiments...and, as says Oppenheimer, *"studies in the history of science can bring some coherence to the general intellectual and cultural life of our time."*

If we try to classify the outstanding works in theoretical physics, we are led roughly to the three following categories:
i) The discovery of formulas from experimental data; among the most famous examples we can mention the discovery of the Kepler laws or that of Balmer's formula[2].
ii) The derivation of a theory from the analysis of "working formulas". As examples, we can mention Bohr's atom theory and Heisenberg's matrix mechanics.
iii) The derivation of a theory from a deep analysis of physical concepts. The most famous example of such a category is the discovery of special relativity. We can also mention the original ideas of de Broglie or those of Yukawa.

Obviously, there are works which belong to more than one category. It is the case of Newton's gravitation theory which was a discovery of types ii (role of Kepler formulas) and iii (the apple and the moon). It is also worthwhile to underline that quantum mechanics was founded in a twofoldway: type ii (Heisenberg) and type iii (de Broglie, Schrödinger).

All these processes of discovery, which were so fruitful in the past, seem now to be given up for mysterious reasons I am not able to explain. A new category was born a few years ago (see superstrings).[3]

In order to try to improve the present quantum theory, we could either try to

---

bibliography given at the end of Chapter 1).

[2] Pertinently, L.C. Biedenharn suggested me to add, as an example, the Planck formula.

[3] There are two attitudes which are difficult to justify in research in physics since, up to now, they never appeared to be fruitful: one consists in trying to introduce a new mathematical structure in physics (let me remind the reader that Einstein did not want to introduce differential geometry for the reason it was a new mathematical structure); the second one consists in believing that any classical object which can be imagined has necessarily its quantum mechanical counterpart in the physical world, a counterpart which is obtained by the trick called quantization, as if quantization was a procedure taught by Nature. A remarkable fact is that these two attitudes interact very well to-day in superstring theory. The fact that it is a very nice theory is not at all an argument in favour of its physical value. Perhaps the most important thing we learn from superstrings is what the good physical theory is not. I could risk a comparison: in a sense, Democritus was more justified in building his atomic conception of matter than the contemporary physicists in introducing the superstring structure.

In order to try to improve the present quantum theory, we could either try to discover the theory which is hidden behind the Feynman graph calculations or to analyze the physical concepts we are familiar with. My feeling is that the first approach requires the talent of outstanding mathematicians. As a modest theoretician, I will concentrate in the present lectures on the analysis of some physical concepts and, more precisely, on the one of *localization* (and *localizability*). I will show the difficulties, and sometimes the contradictions, one encounters in examining this concept and then, I will invite the reader to follow some specific directions which look promising. The contents of my lectures are based on personal reflections, on the readings of great authors and on discussions I had with the mathematician Alain Connes. Without these discussions, I would have much less to say. I would be satisfied if these lectures lead some readers to read or reread some of great master's works. It would be marvellous if one of my readers was able to improve the present analysis and was led to some new promising idea. Obviously, that is my secret hope.

Now, let us enter the subject of our lectures. If I asked you what we call the Schrödinger position operator, you would answer immediately and unanimously: it is X. But what is X?

The question is easy to answer but it takes time to make it precise. We start with the *"ordinary space"* E on which we define the Hilbert space of square integrable functions L^2(more precisely, if f and g are such functions and M a point in E, we do not distinguish between f and g if $f(M) - g(M) = 0$ everywhere except on a subset of zero measure).

Two preliminary remarks:
a) The standard notation for a point in E is not M but x (or r). This is based on the implicit identification of E with R^3 with the aid of an orthonormal Cartesian frame.
b) The measure used in the definition of L^2 is the standard Lebesgue measure in R^3.

Now, X is the operator "multiplication by x", an operator acting on L^2. We see that the identification of E with R^3 is essential. But is this identification justified? That it is a complex question can be shown in decomposing it into three other questions:

Question 1: *Is our space a set?*
Question 2: *If it is a set, is it a continuum?*
Question 3: *If it is a continuum, is it isomorphic to R^3 ?*

What is known is that, in the approximation of classical physics (general relativity excluded), the R^3 structure works very well, both for particles and waves. I do not think that the concept of a classical wave needs any comment. The role of R^3 is quite clear as is the link between waves and the theory of partial differential equations. It is not so, however, for the concept of particle in classical mechanics which is a very vague one. In the mind of the physicists of the classical period, a particle was a small piece of matter, small enough to be identified in a good approximation with a point in E. This implies, of course, that we can also neglect the angular momentum and the corresponding rotation energy of this piece of matter. At the end of the XIXth century, the smallest particles were the atoms and smaller particles were to be discovered. I would like to insist on the fact that a *physical* particle was always thought at that time as a subset of *E* filled with matter. Even when Thomson gave his own model of atom, he thought of a kind of a plum-pudding with electrons instead of raisins. Similarly, the first model for an electron was a sphere of matter uniformly charged. Then, from a classical point of view, a particle is a continuous set of points of matter, each point lying at a point of *E*. A *part*icle lies in a *part* of *E*.

Today, every physicist knows that such models are not acceptable, the particle is a more abstract concept but he does not give up the point structure of *E* . In other words, paradoxically, the concept of space is the same in quantum mechanics and in classical mechanics. If we now examine the problem of the *position* of a particle, we readily see that we have to answer the following fourth question:

Question 4: *Supposing that the answer to the three first questions is yes, is there a canonical way to associate with an elementary particle a point in E ?*

Clearly, this is a question in the classical sense; concerning its quantum mechanical counterpart, I cannot be very precised in such a short introduction, I only want to describe roughly what a physicist has in mind when he thinks of the concept of position; to simplify, suppose that *E* is one-dimensional; we would have to associate with an elementary particle an operator with $\mathbb{R}$ as a spectrum , a real number denoting the coordinate in the classical sense. We readily see that in order of speaking of position in quantum mechanics, we need not only the classical space *E* itself but also a classical frame. The quantum mechanical counterpart of question 4 would be:

Question 4 bis: Supposing that the answer to the three first questions is yes, does exist a canonical measurement (an ideal one) concerning an elementary particle permitting to associate a point in *E*?

It is clear that answering this question affirmatively implies the existence of three commuting operators with $\mathbb{R}$ as a spectrum, one for each coordinate in *E*. This requirement is valid in both relativistic and non relativistic quantum mechanics. A more precise formulation of Question 4 bis has been made by Newton and Wigner; the problem was investigated in a more rigorous and systematic way by Wightman. I will comment about it later on. In this brief introduction, I only want to underline that the Schrödinger position operator was completely satisfactory for a non spinning non relativistic particle. Once a frame in *E* is chosen, the state of the particle is described by a function $\psi(x,y,z)$ of L^2. The Schrödinger position operator is the set X, Y, Z (multiplication by *x, y, z*, respectively). The position measurement is an ideal procedure providing us with three real numbers *x, y* and *z*. To be more realistic, any position measurement is a procedure permitting us to answer the question: *"is the particle in a given portion of E"*? Since *E* is a classical concept, such a measurement corresponds to what we could call a classical question for a quantum object. But it is fair to ask ourselves: *does the quantum particle know anything about E?* This is the main question we will examine in the present volume. For the moment I only want to underline some of the difficulties we encounter when we want to go further in quantum theory.

i) If we have more than one particle, say two, the wave function is defined on $E \times E \times R$ (R for time); the classical interpretation of a wave in the ordinary space E disappears.

ii) When we go from the Schrödinger equation to its relativistic generalization, difficulties arise. Obviously, the space $E \times R$ is replaced by the Minkowski space-time, but *a)* covariance is broken since x, y, z have corresponding observables but not t; *b)* there is no natural generalization of the carrier space $E \times E \times R$ for two particle states; *c)* the potential formalism which works for the Schrödinger equation is rigorously impossible for relativistic equations (no reduced mass, Klein's paradox, etc.).

iii)The Bohr correspondence principle tells us that we must be able to derive classical physics from quantum physics. According to this philosophy, we must be able to define E from quantum physics itself. It is not yet the case to-day. It is perhaps worthwhile to underline that quantization is only a *trick*, a trick permitting to get a true theory from an approximate one; it follows that a good theory must ignore the quantization procedure.

In the next chapter, we will investigate some of the difficulties the physicists encountered in building quantum physics. Many of them disappeared during the construction of quantum field theory. Others were simply forgotten, due to the rapid successes of QED. In the next chapters we will concentrate on the present conceptual contradictions of quantum theory and we will give some proposals to solve them.

Just a word before closing this introduction: I said at the beginning that my purpose is to consider foundations of quantum theory as established. This means that I will not enter the problem of the interpretation of quantum mechanics in these lectures.

CHAPTER 1

HISTORICAL SKETCH ON THE NATURE OF LIGHT

Light is,without doubt, the phenomenon which is at the very origin of the modern revolutions in physics. First, it is the black body radiation which conducted Planck to introduce the notion of *quanta of energy* and the photoelectric effect which led Einstein to propose the notion of *light quanta*.. Second, light signals were extensively used to clarify the notions of space and time and to help Einstein in constructing special relativity. Third, it is the analogy between optical waves and analytical mechanics (through Fermat and Maupertuis principles) which led de Broglie to the proposal of wave mechanics. In contradistinction with these facts, the photon is probably the less understood concept today. As an illustration of this fact, let me mention that the democracy introduced by de Broglie in 1923, in extending the wave-corpuscle duality to all kinds of particles, was destroyed three years later by the probabilistic interpretation of the wave function by Born, since the wave function associated with the photon cannot be given such an interpretation as we will see later.[1]

These are the main reasons why I think it is worthwhile to start these lectures by a brief history of the physical nature of light. This chapter does not pretend to be a complete report on the subject. My only purpose is to analyze some of the many difficulties and contradictions which appeared in the development of modern physics. I suppose the reader to be familiar with the main historical facts; he is strongly invited to refer to the bibliography given at the end of the chapter.

I will divide my historical sketch in three periods, considered as acts of a drama.

Act I (up to 1853)   waves *or* corpuscles?

Act II (1853-1905)   waves!

Act III (since 1905)   waves *and* corpuscles (or, rather, *neither* waves *nor* corpuscles).

---

[1]This renunciation of democracy was made inadvertently. The idea of democracy between all kinds of particles was strongly upheld: it explains why Jordan quantized Schrödinger waves and, later, why the meson was proposed by Yukawa.

ACT I (....-1853)

I do not intend to discuss the well-known fight between wave and particle people. The main actors of this period were Huygens (1629-1695), Newton (1643-1727)[2] and Fresnel (1788-1827). The final point in the dispute was provided by the experiment suggested by Arago (1786-1853)[3] and performed by Foucault (1819-1868) in 1853.[4] The result of the experiment was in favour of the formula c/n for the speed of light in a medium with a refractive index n, that is in favour of the wave theory, and not $c.n$ as predicted by the emission theory. The difference is illustrated by the variational principles

of Fermat (1661) $\delta \int n \, ds = 0$, where $n = n (x, y, z)$

of Maupertuis (1746) $\delta \int v \, ds = 0$, where $mv^2 = 2[E - V(x, y, z)]$

The fight was strong. Let me mention that Poisson, who was defending the corpuscle cause, criticized the wave theory as predicting a stupid fact: *light behind a small circular screen lit by a point like source!* The experiment - a beautiful one - was made: another success of the wave theory!

It is important to underline that there were deep arguments opposed by corpuscles physicists and, at the end of the nineteenth century, although the wave victory was no longer brought into question, the two following conceptual difficulties were still unsolved:
- Aether must be given a very high rigidity to explain high frequencies; such a rigid medium is supposed to fill the vacuum and all transparent materials.
- Why do we have only transverse waves?

These two objections were considered as very serious ones; nevertheless, the physicists accepted the verdict: *light = waves*. Until 1905! As everybody knows, this is an important date, the end of Act two.

---

[2] About the idea of Newton's emission theory, see M. Sachs in [Hoo].

[3] " *Deux points rayonnants placés l'un près de l'autre et sur la même verticale brillent instantanément en face d'un miroir tournant. Les rayons du point supérieur ne peuvent arriver à ce miroir qu'en traversant un tube rempli d'eau; les rayons du second point atteignent la surface réfléchissante, sans avoir rencontré dans leur course d'autre milieu que l'air. Pour fixer les idées, nous supposerons que le miroir, vu de la place que l'observateur occupe, tourne de droite à gauche. Eh bien! si la théorie de l'émission est vraie, si la lumière est une matière, le point le plus élevé semblera à gauche du point inférieur; il paraîtra à sa droite, au contraire, si la lumière résulte des vibrations d'un milieu éthéré...*"(François Arago, quoted in [Cos]).

[4] "*...il ne peut plus subsister le moindre doute sur la véritable valeur de la vitesse de la lumière dans l'espace vide ou dans notre atmosphère. Quant aux vitesses que prend la lumière en pénétrant dans les milieux réfringents, elle n'était donnée que par le calcul, qui, interprétant la réfraction dans le système de l'émission ou dans le système des ondulations, donnait, selon l'hypothèse adoptée, des résultats bien différents. M. Arago, dès l'année 1838, fit le premier sentir l'importance d'une expérience qui, sans même conduire à la mesure exacte des vitesses de la lumière dans les milieux inégalement réfringents, mettrait seulement leur différence en évidence et fixerait, par suite, les physiciens sur la manière d'interpréter la réfraction.*" (Jean-Bernard-Léon Foucault, quoted in [Cos]).

ACT II (1853-1905)

The drama is now played on two separate stages. On the first one, one can see how the physicists were progressively led to reject the aether and see the continual improvement of the *field* notion. The roots of this part of history lie in the works of Ampère (1775-1836), Faraday (1791-1867) and Maxwell (1831-1879). It culminates with the building of special relativity by Einstein (1879-1955). On the second stage, the body radiation, postulated in 1791 by Prévost (1751-1839), is investigated intensively until Planck found the key of the contradictions of the laws of radiation, in providing the physicists with an ansatz: *energy emitted and absorbed by quanta*. The drama culminates on this stage when Einstein proposed to consider light as composed of such quanta! The main events of Act two are sketched in Table 1. It is a remarkable fact that the same man appears as the hero on the two stages and that the two curtains fall simultaneously.

We must underline that during the whole second act, light was considered as a wave phenomenon governed by Fresnel's equations. The fact that Maxwell reinterpreted light as an electromagnetic phenomenon had no effect on its wave nature[5]; the Fresnel vector became the electric vector. From the mathematical point of view, the theory was considered as a perfect one but none explanation was given about the two difficulties concerning the rigidity of the aether and the absence of longitudinal waves. This explains, in my opinion, why Poincaré entitled his book *Théorie mathématique de la lumière* [6]: the theory was not satisfactory from the conceptual point of view; it was *just* a mathematical theory; the nature of light was not really understood, even if, since Maxwell, the physicists knew that electromagnetic waves were synonymous of aether vibrations.

If I mention the Poincaré book, it is to emphasize the similarity of the situation of electrodynamics at that time and the present situation of quantum electrodynamics. In both cases, the theory is able to provide, in principle, all numerical results we need but, in both cases, the concepts used are not satisfactory. In fact, the situation is even worse to-day because we cannot refer to QED as a *mathematical theory* but rather as an *algorithmic* one[7]. In fact, it would be very nice if we were able to discover the mathematics which are behind the Feynman diagrams and renormalization calculations.

---

[5]The aether got one more property: it became a conductor.

[6]The reader could object that such a title is explained by the fact that Poincaré was himself a mathematician and that the book was not intended to penetrate physical aspects. That it is not true can be ilustrated by the fact that the Poincaré book is probably the only textbook in optics where, after the proof of the refraction sinus law for electromagnetic plane waves, a natural physical question is asked and answered, namely: are we permitted to identify a pencil of rays used by an experimentalist (to check the sinus law) with a plane wave? The Poincaré conclusion is: yes, provided the width of the pencil is at least ten times the wavelength.

[7]*"Working with the present foudations [of quantum mechanics], people have done an awful lot of work in making applications in which they can find rules for discarding the infinities. But these rules, even*

	Stage 1: *Aether disappears*	Stage 2: *Towards a gaz of photons*
1859		Radiation equilibrium, Kirchhoff(1824-1887)
1862	Displacement current	
1865	E.M.waves have speed c, Maxwell(1831-1879)	
1879		Stefan law, Stefan(1835-1893)
1884		Radiation pressure, Boltzmann (1844-1906)
1887	Michelson experiment, Michelson(1852-1931)	
1888	Electric waves, Hertz(1857-1894)	
1893		Displacement law, Wien (1864-1928)
1894	Aether? Poincaré(1854-1912)[8]	
1896		Wien law
1900	Light carries a momentum, Poincaré and Lorentz(1853-1928)	Planck law, Planck(1858-1947)
1905	Special relativity Einstein(1879-1955)	Light is made of quanta, Einstein[9]

Table 1

though they may lead to results in agreement with observation, are artificial rules, and I just cannot accept that the present foundations are correct" [Dir, p.20].

[8]It is possible to see the progressive disappearance of the aether in comparing the two successive editions of the *Théorie mathématique de la lumière* by Poincaré. In the first one, the first chapter is devoted to the theory of elasticity. Two vectors are introduced which are interpreted in the second chapter as the electric and magnetic vectors. Two years later, in the second edition, Poincaré starts directly with Maxwell equations and *then* shows the analogy of the electromagnetic field with elasticity. However, he did not discard the aether explicitly.

[9]I do not say "a gaz of particles" because the *photon* (named by Lewis in 1926) was not recognized completely at that time as a particle (it will be given a momentum in 1917 by Einstein). Einstein always referred to the photon as a *light quantum*. Even to-day, there are physicists who are not considering it as a particle but as a quantum of a gauge field. Moreover, if we want to use the expression "gaz of particles", it would be better to be more precise and to say "a gaz of indistinguishable particles". As says Schrödinger:"....*The recognition that the thing which has always been called a particle and, on the strength of habit, is still called by some such name is, whatever it may be, certainly* not *an individually identifiable entity*" [Sch1, vol.3, p.702].

Before going to Act three, we have to emphasize the following remarkable fact: in the same year 1905, at the moment where Einstein was inviting the physicists to give up the aether[10] which was at the origin of the two difficulties of the wave theory of light, he himself was bringing into question the wave theory of light...Let us also emphasize that giving up the aether transformed fields into very respectable objects not requiring any material support. This provided Maxwell theory with a sublime character which explains probably why Maxwell equations were kept as a foundation ingredient of quantum electrodynamics.[11]

ACT III (1905-....)

It took *twenty years* for Einstein to convince his colleagues of the existence of light quanta. This does not mean that he was rejecting the classical theory of Maxwell. He wrote in 1924 *"There are therefore now two theories of light, both indispensable and - as one must admit to-day in spite of twenty years of tremendous effort on the part of theoretical physicists - without any logical connection"* (quoted in [Dre, p.199]). For us who are so familiar with the quantum explanation of the photoelectric effect, such a delay is difficult to believe. The fact is that the first photon to be "seen" was in an experiment performed by Compton and Simon in 1925 (three years after the discovery of the Compton effect) where it was possible to interpret , in a Wilson chamber, the travelling of a photon between two points corresponding to a double scattering. It was really this experiment which persuaded the physicists of the reality of light quanta. There was a good reason to be reluctant to such objects: they were unable to explain interferences; no possibility was seen to conciliate the continuous character of waves and the discreteness of light quanta. That explains why Bohr, Kramers and Slater proposed in 1922 a theory *avoiding light quanta*. In his Nobel lecture in 1922, almost at the time of Compton's paper, Bohr declared himself against light quanta (Table 2).

1925 can be considered as the official year of the acceptance of the light quantum; paradoxically, *it was two years after the birth of quantum mechanics!* It seems natural to consider this period (1923-25) as separating the first two scenes of Act III. For our purpose, I do not intend to develop extensively this act. The main facts which occured before the death of Einstein are sketched in Tables 2, 3 and 4. It is important to underline the strong (and long!) drama of Einstein who was confessing almost at the end of his life that he never really understood what was a light quantum although he spent twenty years to convince his colleagues of the reality of the photon.

---

[10]Not exactly; Einstein underlines in a talk given at the University of Leyden (May, 1920) that relativity does not obliges us to reject the aether: *"Une réflexion plus attentive nous apprend pourtant que cette négation de l'éther n'est pas nécessairement exigée par le principe de la relativité restreinte. On peut admettre l'existence de l'éther, mais il faut alors renoncer à lui attribuer un état de mouvement déterminé, c'est-à-dire il faut le dépouiller par l'abstraction de son dernier caractère mécanique, que Lorentz lui a encore laissé"*[Ein].

[11]However, the mathematical beauty of Maxwell theory is partially obscured in quantum electrodynamics.

Scene 1: <u>Towards the photon (1905-1923)</u>

1905	The photoelectric explained (Einstein).
1913	The Bohr atom.
1917	The light quantum has a momentum (Einstein).
1918	*"I have no more any doubt about the reality of quanta in radiation, although I am still alone with this conviction "* (Einstein to Besso).
1922	The Compton effect .
1922	Bohr-Slater-Kramers' theory. *"In spite of its heuristical values...[Einstein's] hypothesis of light quanta is irreconcilable with so-called interference phenomena, is not able to throw light on the nature of radiation".* (Bohr's Nobel lecture).
1925	The Compton-Simon experiment: it is the birth of the photon (the word itself is introduced by Lewis in 1926).

Table 2

Scene 2: <u>The birth of quantum mechanics.</u>

1923	De Broglie's waves.
1922/4	Light quanta statistics (de Broglie, Bose, Einstein).
1924	The exclusion principle (Pauli).
1925	The spin of the electron (Uhlenbeck and Goudsmit).
1925/6	Matrix mechanics (Heisenberg, Born, Jordan).
1926	The Schrödinger equation.
	The Fermi-Dirac statistics.
	Born's statistical interpretation.

Table 3

	Scene 3: Quantum mechanics and relativity.
1928	The Dirac equation as a relativistic substitute of the Schrödinger equation (successes: $g=2$, Thomas' $1/2$ factor; difficulty of the negative energy states).
1930	Schrödinger's proposal for solving the problem of negative energy states: a new relativistic position operator for the electron.
1931	Dirac suggests the existence of the positron[12].
1932	The positron is discovered (Anderson).
1939	Elementary systems defined with the aid of the Poincaré group (Wigner).
1949	The photon is not localizable (Newton and Wigner).
1951	*"All the fifty years of conscious brooding have brought me no closer to the answer to the question "what are light quanta?" Of course, to-day, every rascal thinks he knows the answer, but he is deluding himself"*(Einstein to Besso, quoted in [Stu], p.332).[13]

Table 4

It seems that, even after the discovery of the Compton effect, Bohr was still reluctant to accept the concept of light quantum. *"As Ehrenfest reported to Einstein at the beginning of 1922, Bohr was much more inclined to abandon the energy-impulse-conservation principle for the elementary processes 'than to shift the blame on the aether'. Bohr was not even willing to understand the Compton effect as supporting the particle point of view"* (Karl Von Meyenn in [Lah]). *According to the same source, it is only in 1925 that Bohr accepted the idea of light quantum and give up the theory he had built with Kramers and Slater: "Now we can do nothing else but to painlessly erase all traces of our revolutionary attempt"*[Lah][14].

---

[12]About the non obvious character of that prediction, see [Dir].

[13]Franck says in his Einstein's biography that *"[Einstein] studied all works of the great physicists to the purpose of finding whether they could contribute to the solution of this problem concerning the nature of light"*.

[14]See [Dre] where the historical role of the Bohr-Kramers-Slater theory is extensively studied. According to Dresden, almost all physicists were ready to adopt it, some of them with much enthousiasm (Born), others with reservations (Heisenberg). The main objections came from Einstein and Pauli. Biedenharn us right to underline (private letter):*" The BSK theory had one very new and correct point: the idea of virtual states which was developed by Slater. It is ironic that Slater's best physics contribution got buried in the difficulties introduced by his famous collaborators"* .

It is hard to underestimate the historical role played by the Compton experiment. As we underlined in Table 2, it was a quite impressive quantitative proof that the light quantum behaves like a particle with its own energy-momentum; it was really the birth of a new particle. Then it was time for de Broglie to put on the same footing the photon and the other particles. The de Broglie extension of the wave-particle duality was a precursor idea for the Yukawa extension of the field-particle duality (the meson). Such a democracy[15] is also present in the Wigner paper on elementary systems described with the aid of irreducible representations of the Poincaré group. But, as we already mentioned at the beginning of the present chapter, this democracy was innocently destroyed in 1926 by the Born statistical interpretation of the wave function.[16]

To sum up this historical introduction, I would say that what I had tried to do is to persuade the reader that the 1951 quoted sentence of Einstein deserves looking into, despite Bohr's efforts. Chapters 2, 3 and 4 will be devoted to many of the difficulties of quantum theory in which the photon concept plays a central role. Chapter 5 will describe an attempt to solve one difficulty about the photon, namely the problem of its localization. We will see that it will help to clarify the notion of complementarity of Bohr. In Chapter 6, we investigate some consequences for the other particles and will show how the spin-orbit coupling appears as a by-product. Finally, we examine in Chapter 8, which changes could be expected for quantum field theory and why we have to give up Minkowski space-time.

Bibliography

[Be.E]   Einstein-Besso, *Correspondance* (1903-1955), (Hermann, Paris, 1972)

[Bo.E]   M. Born, *The Born-Einstein Letters* (Walker, New-York, 1971).

[Bro]   L. de Broglie, *Les incertitudes d'Heisenberg et l'interprétation probabiliste de la mécanique ondulatoire* (Gauthier-Villars, Paris, 1982).

[Cos]   M. Cosmovici, *L'évolution de la physique au XIX^{e}siècle* (Larousse, Paris,1914).

---

[15]One of the main fighters in favour of the democracy between all particles was Jordan. *"He was probably the first to be convinced of the need of quantizing the electron wave function and of abolishing the distinction between matter and energy; matter as light could be created and annihilated "*[Dar].

[16]This was revealed - but not made explicit - by a result of the Newton and Wigner paper of 1949: the photon being not localizable, its wave function cannot be given the Born interpretation! About this contradiction, I cannot resist to the temptation to quote Slater [Pri, p.20]: *"When I went to Copenhagen at the end of 1923,...Bohr had not yet brought himself to the point where he would admit the existence of corpuscular photons along with the waves of light. My proposal to Bohr and Kramers had been a straightforward probability connection between the waves and photons: the intensity of the continuous radiation field, at any point of space, was taken to determine the probability of finding photons at that point. Bohr objected to the photons so strongly that it was obvious that I would have a fight on my hands if I insisted on them, though I never doubted their existence...A short time later, the work of Bothe and Geiger (1925)...[forced] Bohr to give up his opposition to the photons...Bohr himself, by 1926, wrote me regretting that he had opposed me in 1923".* As says Pais:*"The history of science is full of gentle irony".*

[Dar]   O. Darrigol, Historical Studies in the Physical and Biological Sciences, *16*, 197 (1986).

[Dir]   P.A.M. Dirac, *Directions in Physics* (John Wiley and Sons, New York, 1978).

[Dre]   M. Dresden, *H.A. Kramers, Between Tradition and Revolution* (Springer-Verlag, New York, 1987).

[Ein]   Einstein, A. *Réflexions sur l'électrodynamique, l'éther, la géométrie et la relativité* (Gauthier-Villars, Paris, 1972).

[F.W]   M Fierz and V.F.Weisskopf (Eds) *Theoretical Physics in the XXth century*(Inters.Publ., New York, 1960).

[Fra]   P. Frank, *Einstein, his Life and Times* (Knopf, New York,1953).

[Haa]   D. Ter Haar, *The Old Quantum Theory* (Pergamon Press, Oxford, 1967).

[Hei1]   W. Heisenberg, *The Physical Principles of the Quantum Theory* (Dover, New York, 1930).

[Hei2]   W. Heisenberg, *Physics and Philosophy* (Harper, New York, 1962).

[Hei3]   W.Heisenberg, *Philosophical Problems of Quantum Physics* (Woodbridge, C.T.Ox Bow, 1979).

[Her]   A. Hermann, *The Genesis of Quantum Theory* (The M.I.T. Press, Cambridge, Massachussets, 1971).

[Hof]   B. Hoffmann, *The Strange Theory of the Quantum* (Dover, New York, 1947).

[Hoo]   C.A. Hooker (Ed.), *Contemporary Research in the Foundations and Philosophy of Quantum Theory* (Reidel, Dordrecht-Holland, Boston, 1973).

[Jam1]   M. Jammer, *The Conceptual Development of Quantum Mechanics* (M Graw Hill, New York, 1966).

[Jam2]   M. Jammer, *The Philosophy of Quantum Mechanics* (Wiley, New York, 1974).

[Lah]   P. Lahti and P.Mittelstaedt, Eds, *Symposium on the Foundation of Modern Physics* (World Scientific, Singapore, 1985).

[Lau]   M. Von Laue, *History of Physics* (Academic Press, New York, 1950).

[Pai1]   A. Pais, *Subtle is the Lord* (Clarendon Press, Oxford, 1982).

[Pai2]   A. Pais, *Inward Bound* (Clarendon Press, Oxford, 1986).

[Poi]   H. Poincaré, *Théorie mathématique de la lumière* (Georges Carré, Paris, 1892).

[Pri]   W.C. Price et al. (Eds), *Wave Mechanics, the First Fifty Years* (London, Butterworths, 1977).

[Sch1]   E. Schrödinger, *Collected Papers*, (Austrian Academy of Science, Vienna, 1984).

[Sch2]   E. Schrödinger, *Letters on Wave Mechanics* (Philosophical Library, New York, 1967).

[Schw1]   J. Schwinger, *Quantum Electrodynamics* (Dover, New York, 1958).

[Schw2]   J. Schwinger, *Particles, Sources and Fields* (Addison-Wesley, Reading, Massachussets, 1970-73).

[Stu]   R.H. Stuewer, *The Compton Effect* (Science History Publications, New York, 1975).

[Wig2]   E.P. Wigner, *Symmetries and Reflections* (Indiana University Press, Bloomington, 1967).

[Wig3]   E.P. Wigner, Int. Jour. Theor. Phys. *25*, 467 (1986).

CHAPTER 2

THE CORRESPONDENCE PRINCIPLE,
THE WAVE-CORPUSCLE DUALITY, THE COMPLEMENTARITY
PRINCIPLE AND THE SLIT EXPERIMENT.

In the present chapter, I intend to make some comments about two important general principles which are commonly considered as essential to understand how to use quantum mechanics. I mean the correspondence principle and the complementary principle. In the following, the reader is supposed to be familiar with what is written about them in standard textbooks. Although these principles are considered as fundamental from the point of view of philosophy, my purpose is not at all philosophical. What I want is to examine them independently of their historical context, as if they were just proposed. Let us examine first the *correspondence principle*. In contradistinction with what is usually taught in the majority of textbooks, many physicists are very prudent, even perplexed, in discussing the value of this principle. Pais is right to say that *"It takes artistry to make practical use of the correspondence principle"* ([Pai2], p.247). Among many opinions, let us quote the following judgement: *"It is difficult to explain in what [the principle] consists, because it cannot be expressed in exact quantitative laws, and it is, on this account, also difficult to apply. In Bohr's hands it has been extraordinary fruitful in the most varied fields; while other more definite and more easily applicable rules of guidance have indeed given important results in individual cases, they have shown their limitations by failing in other cases"*[K.H].

The correspondence principle may be given such a general meaning that it becomes completely useless. This is the case if we formulate it in the following way: given a theory, there must exists a procedure to obtain an approximate theory by supposing that some fundamental constant is given a zero or infinite value. Such a statement is unfortunately meaningless. To prove it, I will take the following example: make $c = \infty$ in the theory of special relativity. One of the most fundamental formulas of this theory is

$$E^2 - P^2 c^2 = M^2 c^4 \qquad (2,1)$$

and when c is going to infinity, we see that the total energy of a particle becomes infinite. This is the expected result, since the rest energy Mc^2 becomes infinite together with c.[1] Now, divide Eq.(2,1) by c^2 and take the limit. We obtain the *unexpected* result

---

[1] The reader would be tempted to say that this result could be added to classical physics without damage. I do not agree. If the total energy of any system is infinite, what would happen to the classical principle of energy conservation?

that P also goes to infinity! Another "stupid" result is obtained by dividing (2,1) by c^4 before taking the limit. It seems amusing to underline that, if we subtract the infinity Mc^2 from the infinite energy, we are left with a concrete finite quantity, namely the kinetic energy. This is certainly the simplest example where a theory needs subtracting an infinite number from another infinite number to obtain an observable quantity. Is it an indication that a theory which has to subtract infinities is necessarily an approximate theory?

This example proves that, to be correct, a correspondence principle must tell us which formulas must be transformed. Obviously, these formulas cannot be chosen arbitrarily, they must lead us to a *coherent family of formulas* because transforming one formula could contradict another transformed formula.

There is much more to say about that. Physics is not just a set of formulas; it is a science of Nature and any correspondence principle must take this character into consideration. Indeed, when we are interested in a correspondence principle, we have in mind a set of two theories, say A and B, we want to compare. Moreover, there is a dissymmetry between the two theories, for instance B is the approximation of A. Therefore, we do know that B is a wrong theory, or, to say the things in a less drastic way, there are physical situations where B gives completely wrong predictions and other ones where the predictions are acceptable. As a consequence, a correspondence principle to be useful must state explicitly under which physical conditions theory B can be applied. Let us take again the example of special relativity. It has been shown by Lévy-Leblond that Newtonian mechanics is obtained from special relativity not only in taking the limit c going to infinity in specific formulas but also in restricting ourselves to experiments which involve small space intervals compared to time intervals [Lév1]. More precisely, the Lévy-Leblond statement is that the Poincaré group, the kinematical group of special relativity can give by *contraction* [2] (c going to infinity) two distinct groups, the Galilei group and the Carroll group.[3] As Lévy-Leblond proved, the two groups are kinematical groups associated *with different physical approximations*.[4] Theories A and B associated with *this* correspondence principle are relativistic and non relativistic *mechanics* but we could be more ambitious: we could ask for a correspondence principle which permits to go from special relativity to pre-Maxwell physics (Newtonian mechanics + pre-Maxwell electrodynamics). Obviously, more laws would be involved and, consequently, more experimental situations. The reader is referred to [Lév2] and to [B.K] for this problem. Let us underline our conclusion. A correspondence principle concerns:

a) a consistent set of formulas of the best theory A describing a given domain of physics,

b) a given fundamental constant appearing in this set of formulas and the value (zero or infinity) it is given to get an approximate theory (Theory B),

c) the experimental conditions for the validity of Theory B.

In the present volume, we are interested in quantum mechanics and the associated

---

[2] a concept introduced by Inönu and Wigner [I.W]. See also [Seg] and [Sal].

[3] named in this way by Lévy-Leblond, after Lewis Carroll, the author of *Alice in Wonderland*.

[4] Obviously, nothing forbids us to consider seriously a world governed by the approximate theory. To know more about the physical meaning of contraction, see [B.L].

correspondence principle involves traditionnally the Planck constant. The reader is invited to play the above game and examine "stupid" proposals as the one which consists in replacing h by zero in the Schrödinger equation. I hope that the reader is now convinced that, before stating a principle of this kind, we have to select in quantum mechanics a consistent set of formulas where we will replace h by zero. We will not look for such a set. We have still the two following open questions: i) how large is the domain of physics we are interested in? and ii) what kind of experiments are permitted in the approximated theory? These two questions are never explicitly answered but let us try to say what physicists have in mind. First, it is clear that they cannot think of an approximate theory (Theory B) which is not self-consistent; in particular, B cannot be the combination of Maxwell electrodynamics and Newtonian mechanics since these two theories are not compatible. Among many possibilities, the correspondence principle under consideration may concern one of the three following possible links:

a) QED to Maxwell theory,
b) non-relativistic quantum physics to pre-Maxwell physics,
c) non-relativistic quantum mechanics to Newtonian mechanics.

The two last cases are quite close because theory A has no room for light quanta and, consequently, there is no room for electromagnetic waves in the approximate theory. It follows that, from our point of view (where the photon is considered as the key of quantum physics), the correspondence principle is of very small interest in these two cases. We are left with case a. Unfortunately, in this case, it seems that the corresponding problem is a pure academic one: I do not know any attempt to derive Maxwell theory from QED.[5]

I do not intend to investigate these problems in more details; I would like to underline that the questions they are asking are of *historical* rather than *physical* order. Let us see why. Going from quantum theory to classical physics[6] suppose that the particle aspect of the photon disappears together with the wave aspect of the electron. Why would it be more interesting for *physics* to consider this limit rather than the opposite, namely, keeping the photon and giving up the electron as a particle? Why do not we consider the wave limit of QED, both for the Maxwell and the electron fields? It is worthwhile to quote a sentence of Jordan written in 1925:"*We can understand the elementary scattering process not only as the scattering of light waves by material corpuscles but also as the scattering of matter waves by corpuscular light quanta*"(quoted in [Dar, p.219]). We have probably quite many approximate theories of QED...

As a conclusion, I would say that I do not deny the value of correspondence principles, provided they are investigated in the framework of physics (in both theoretical and experimental aspects) or in their historical role, but much more work is needed to state them in an acceptable way.

Let us now examine the wave-corpuscle duality and the *complementarity principle*. The author of this principle is Bohr. This is not surprising; as we underlined in the previous chapter, Bohr was reluctant to accept light quanta, essentially because it seemed

---

[5]As a joke, what about the "exercise": derive geometrical optics from quantum chromodynamics?
[6]It is not clear if, in the historical context, the correspondence principle implies the equation: *classical physics = Newtonian mechanics + Maxwell electrodynamics* This is a bastard theory.

impossible to conciliate the *continuity* of Maxwell equations with the *discreteness* of light. As we saw, he tried, with Kramers and Slater to conciliate the continuity of Maxwell equations with the discreteness of emission and absorption. In other words, he was ready to accept Planck's quanta, not the ones of Einstein. It took time for him to recognize the validity of Einstein's concept. It is obvious that accepting light quanta was not enough to solve the contradiction. The complementarity principle was a way of accepting both Fresnel-Maxwell and Planck-Einstein legacies and *to go on*, instead of adopting the critical attitude of de Broglie[7], Schrödinger or Einstein. My opinion is that these three great physicists were right in principle, right in that *there was really a difficulty to solve*; in a sense, Bohr's attitude was *pragmatic*, his principle was just an *expedient* permitting him to go further.

What the complementarity principle asserts, in its simplest form, is that the two aspects of light, namely wave and quanta, are complementary; that means that there are experiments for which the interpretation is ondulatory and others for which it is corpuscular. Obviously, this principle was extended to matter after de Broglie's wave mechanics and Davisson-Germer's experiment. Instead of de Broglie's and Schrödinger's efforts to conciliate the two aspects, Bohr stated his principle according to which, depending of the experiment, only one aspect is present.

I would like to present here two objections to Bohr's idea. First, there are situations where the two aspects compete to interpret a phenomenon. For instance, the Doppler effect has a wave *and* a corpuscular explanation. Second, given two experiments, say A_0, an ondulatory one, and A_1, a corpuscular one, it could exist a continuous set of experiments $A(x)$ such that $A(0) = A_0$ and $A(1) = A_1$. Then, where is the frontier in this set which separate the wave experiments from the corpuscle ones?

Let us illustrate such a situation by a famous thought experiment, namely the electron slit experiment, as it is described by Feynman in his lectures [Fey]. I will not describe it, because every physicist knows it very well. I will concentrate on a part of this experiment, namely the *continuous* way it is possible to make the interference pattern to appear or to disappear progressively. It is well accepted that the fringes disappear completely if we try to know which slit each electron went through, with the aid of an *intense* light source. At this stage, a natural question arises: what happens if we diminish continuously the intensity of light? Everybody knows the answer: the interference pattern reappears progressively. The important thing is that, to explain why, we need two kinds of explanations, depending on the way the radiation is becoming less and less intense: *a*) if the wave length is fixed, then we must diminish the number of photons emitted per second; therefore the probability of a photon-electron scattering decreases which means that there are electrons which are not seen; these electrons interfere; *b*) suppose now that we make the wavelength increasing and keep fixed the flux of photons; all the electrons are seen but in a worse and worse way, because diffraction of light does not permit us to determine the exact position of the electron (the optical image of a point is not a point, but a spot the dimensions of which increase with the wavelength); therefore, when the wavelength increases, we cannot say, for more and more electrons, through which slit they passed. That is the reason why the fringes are reappearing progressively.

---

[7]In fact, de Broglie rallied for a while the so called *Copenhagen school*. It was his "pragmatic period".

It is clear that in the first case, the explanation is of corpuscular type but, in the second it is of ondulatory type. You could call on the Bohr complementarity principle in saying that we considered two distinct experiments, but this is wrong for the following reason. Let us denote by v the frequency of the light source and by n the number of photons emitted per second. With each point in the plane (v, n) is associated one experimental setup. Where is the frontier in this plane between the wave and the corpuscle experiments? If you think that there is no frontier, it implies that the two aspects are simultaneously present, at least in some domain; it follows that we really have to conciliate the two, as claimed in particular by Einstein, de Broglie and Schrödinger. Moreover, let me emphasize another inconsistency in this thought experiment: *although we are interested in an interference experiment, that is to say an ondulatory property of the electrons, we explain the reappearing of the interference pattern in both cases a and b in considering electrons as particles!*

As a teacher, I must confess that I am always embarrassed in teaching a principle which is presented as essential in all textbooks but I am reluctant to accept; I must also testify that it is hard for my students to understand it and they are very sorry of that. I mentioned the names of Einstein, Schrödinger and de Broglie; it is natural to know the opinion of Heisenberg; let me quote him: *"The concept of complementarity introduced by Bohr... has encouraged the physicist to use an ambiguous rather than an unambiguous language, to use the classical concepts in a somewhat vague manner in conformity with the principle of uncertainty, to apply alternatively different classical concepts which would lead to contradictions if used simultaneously... When this vague and unsystematic use of the language leads into difficulties, the physicist has to withdraw into the mathematical scheme and its unambiguous correlation with the experimental facts."* [Hei2]. Heisenberg took refuge in the mathematical formulation of quantum theory to escape the difficulty. It is a surprising attitude for the physicist who stated the uncertainty principle.

Let me conclude: if photons exist - and I think that everybody believes that they do exist - there must be a photon explanation of experiment *b*. It is one of the aim of these lectures to provide such an explanation.[8]

Bibliography.

[B.K]   H. Bacry and J. Kubar-André, Int. Jour. Theor. Phys. 7, 39 (1973).
[B.L]   H. Bacry and J.-M. Lévy-Leblond, J. Math. Phys. 9, 1605 (1968).
[Coh]   L. Cohen, in C.A. Hooker, Ed., *Contemporary Research in the Foundations and Philosophy of Quantum Mechanics* (Reidel, Dordrecht-Holland, Boston, 1973).

---

[8]See Chapter 5.

[Dar]   O. Darrigol, Historical Studies in the Physical and Biological Sciences, *16*, 197 (1986).

[Fey]   R.P. Feynman, R.B. Leighton, M. Sands, *The Feynman Lectures in Physics*, vol.III, Chapter 1 (Addison-Wesley, New York, 1965).

[Hei2]   W. Heisenberg, *Physics and Philosophy* (Harper, New York, 1962).

[I.W]   E. Inönu and E.P. Wigner, Nuovo Cimento *9,* 705 (1952).

[K.H]   H.A. Kramers and H.Holst, *The Atom and the Bohr Theory of its Structure,*(Knopf, New York 1923).

[Lév1]   J.-M. Lévy-Leblond, Ann. Inst. Henri Poincaré *3,* 1 (1965).

[Lév2]   J.-M. Lévy-Leblond, Comm. Math. Phys. *6,* 286 (1967).

[Lév3]   J.-M. Lévy-Leblond, Bulletin de la Société Française de Physique, *14,* Encart pédagogique 1, 1973.

[Pai2]   A. Pais, *Inward Bound* (Clarendon Press, Oxford, 1986).

[Sch1]   E. Schrödinger, *Collected Papers* (Austrian Academy of Science, Vienna, 1984).

[Sal]   E. Saletan, J. Math. Phys. *2,*1 (1961).

[Seg]   I.E. Segal, Duke Math.J. *18,* 221 (1951).

CHAPTER 3

THE SPIN QUANTIZATION PROBLEM.

The quantization problem is generally presented as follows: given a classical observable, that is a function on phase space, what is its quantum equivalent, in the Heisenberg sense? Before entering this problem from the conceptual point of view, it is important to underline that it is the inverse problem which has a physical meaning. Indeed, if we believe in quantum physics as a better theory than classical mechanics, the natural question is to understand how we can define classical observables from quantum ones, since it is natural to be able to derive an approximate theory from the exact one rather than the converse. We could also say that quantization must be considered, at best, as a trick[1] which permits to guess from a non satisfactory theory what could be the right one. This trick is unknown from Nature; it cannot be considered as a law. The inverse of the quantization procedure has a name; it is the *correspondence principle* expressed in the following general form: there must exist a procedure to get from quantum theory a theory which contains all results of classical mechanics.[2]

This small discussion has an important consequence. When quantum mechanics was discovered, the only classical observables which were known were the functions on phase space and it was believed that with each such observable could be associated a unique quantum equivalent; such a one-to-one correspondence was broken when spin was discovered and it was said that spin components were pure quantum observables without classical equivalent. Obviously, if we are referring to standard classical mechanics of Newton, Hamilton, Lagrange, Poisson, etc., it is just a fact. This does not mean that it does not exist an intermediate theory which is *classical*, in that *a)* there is no quantum number, *b)* it has spin variables and *c)* it is equivalent to Newtonian mechanics when all spins are zero. As we will see, such a theory exists. Let us denote by C' this theory, by C the Newtonian one and by Q the quantum theory. We have the following chain of approximations:

$$Q \underset{1}{\to} C' \underset{2}{\to} C \qquad (3,1)$$

The arrow 1 corresponds to the disappearing of any quantum number; the second arrow

---

[1] About quantization, Dirac says that *"it was a good description to say that it was a game, a very interesting game one could play"* [Dir,p.7].
[2] It is important to underline that the correspondence in question is the one relating the quantum theory of massive particles to the classical theory of these particles. See also the beginning of Chapter 2.

describes the vanishing of all spins. There exists a quantization procedure (the Kostant-Souriau quantization) which permits to go from C' to Q and which reduces to the usual canonical quantization when all spins vanish. Obviously the first arrow does not say that Planck's constant goes to zero. If it did, spins would disappear immediately; it only says that all observables which have dimensions of an action can take *any* real value.

Before deciding that the spin components were quantum observables without classical counterparts, the physicists tried hard to find classical models for spinning particles. The first model was the *spinning top* which is a system with three degrees of freedom. By quantizing it in the canonical way, one arrives at a quantum system with *too many quantum numbers*. In particular, each angular momentum L has $(2L+1)^2$ states. In other words, instead of a finite number of components for the wave function, as it is the case in the Pauli-Schrödinger or Dirac equation, the quantization of the top was providing the physicists with an infinite number of components. The spinning top cannot be considered as a classical model for an *elementary* spinning particle. The first to consider an acceptable classical model for spin was Kramers. I will describe it later on.

Spin was not the only difficulty encountered in the quantization problem. The other one is due to relativity; the fact that the position of a particle is a quantum operator and time just a parameter is hard to accept in relativity where covariance requires to put position and time on the same footing. Usually, one escapes this objection by saying that the right quantum theory is relativistic quantum field theory (QFT) but i) it is a simple matter to restrict QFT to one particle states and ii) it is not clear how we derive non relativistic quantum mechanics from QFT. Another pertinent question arises: *why do we think to be right in using canonical quantization for quantizing fields, a procedure which is unable to quantize spin?*

In standard textbooks, quantum theory is introduced in successive steps: i) non relativistic spinless particles, ii) spin, iii) relativistic equations. The first step is characterized by the following facts: *a*) There are roughly two equivalent approaches, the wave function one and the Heisenberg one; *b*) the Heisenberg approach is related to classical mechanics through quantization; *c*) interactions between systems are described by Hamiltonians. In the second step, property *b* disappears; in the third step, we are left with a half of property *a*: the wave equation approach. A great improvement was made about *free* elementary systems by Wigner in 1939 when he defined the Hilbert space of the states of a *free* relativistic elementary system as the carrier space of a projective unitary irreducible representation (unirrep) of the Poincaré group. The Poincaré group to be considered for an elementary particle depends on the particle. Generally, it is the two-sheeted group (the one with parity) except for particles only involved in weak interactions (neutrinos) for which it is the connected group only. Each unirrep is characterized by mass and spin. The advantage of the Wigner approach is not only to have a Hilbert space on which we know how to perform Poincaré transformations, but also to provide us with natural quantum observables, the *momenta* . The energy H (free Hamiltonian) is associated with the time translations, the linear momentum P with space translations and the angular momentum J with the rotations. Since time translations commute with both space translations and rotations, we have

$$[H, P] = 0, \quad [H, J] = 0 \text{ and, trivially, } [H, H] = 0$$

which imply that P, J and H are *constants of the motion*. Obviously, any function of these observables are also constants of the motion, but P, J and H are moreover *additive observables* [3] which means the following: consider a system composed of non interacting elementary systems. Its states span a Hilbert space which is the direct product of the Hilbert spaces associated with each elementary system. The action of the Poincaré group $\wp$ is given by the direct product of representations and the corresponding momenta are obtained by adding the elementary momenta.

Very often, when one refers to the Wigner work, it is to say that what Wigner did was to associate with each kind of elementary particle a unirrep of the Poincaré group. It is important to underline that Wigner did not refer to elementary particles but to *elementary systems,* which constitute a larger class of objects. For instance, a hydrogen atom in its fundamental state is an elementary system with a given mass (a little bit less than the sum of the proton and electron masses) and spin zero. The set of all states of the hydrogen atom form a representation space for a reducible representation of $\wp$ but, unfortunately, group theory says nothing about the elementary particles (proton and electron) which compose the atom. *The reason is that these particles are interacting.* The Poincaré group is unable to associate with a non isolated system any momentum or energy observable. But it is a marvellous tool for analyzing a reaction with the aid of the momenta additivity of the ingoing and outgoing *elementary systems.*

It is not a simple matter to describe interactions between particles in classical special relativity. The things are more complicated in quantum mechanics since the intermediate fields carry quanta and, for that reason, we usually say that quantum field theory is the natural framework for quantum relativity. Strictly speaking, this is not true because the quanta of the intermediate fields may as well be called particles; therefore, it is legitimate (although not orthodox) to say that quantum relativity describes particles interacting *without fields.* We will come back to this aspect in another chapter.

At the beginning of this chapter, I said that there was a "classical theory" of elementary particles, a theory denoted by C' in Eq.(3,1),which was taking into account spin variables. C' has also the advantage of describing massless particles with or without helicity. Moreover, by quantizing it in a suitable manner, we get the Wigner theory of elementary particles. This concept of a classical spinning particle was proposed by Bacry in 1967, then, independently, by Arens and was extensively studied by Souriau. The method of quantization used in this scheme is due to Kostant and Souriau (independent works). The essential idea of C' can be described as follows.

According to Wigner, an elementary particle is described by a Hilbert space on which the Poincaré group acts irreducibly. This means that any state of this Hilbert space is

---

[3] The momenta are not only constants of the motion and additive observables, they are also conserved quantities; this means that they are constants of the motion for any isolated system (example: the electric charge is a conserved quantity). Obviously, any function of the momenta is also a conserved quantity. It follows, in particular that the so-called invariant mass defined by $m^2 = H^2 - p^2$ is also a conserved quantity. Very often the mass is said not to be conserved; in fact what is understood is that it is not additive.

cyclic or, in other words, that any state is obtained from any other one by combining the three following types of operations:

a) performing Poincaré transformations on the initial state,

b) superposing linearly the states obtained in *a*,

c) taking the limit of a Cauchy sequence of states obtained by operations *a* and *b*.

In order to find an analogous definition for a classical elementary particle, it is natural to drop operation *b* (the superposition principle is ignored in classical mechanics). Clearly, operation *a* must be kept since it is clear that any Poincaré transformation must transform a classical state of a particle into another state of the same particle. This means that we require the transitive action of the Poincaré group on the set of states *S*. In group theoretical language, this property reads:*the space of states is a homogeneous space S of the Poincaré group*. Clearly, *S* is a finite-dimensional manifold and, consequently, no requirement of type *c* is needed.

In the quantum case, the set of states has a structure, that of a Hilbert space[4] and $\wp$ acts unitarily, that is, in preserving this structure. In the classical counterpart, it is natural to require that the Poincaré group acts by *canonical transformations*, which implies a phase space structure of the homogeneous space *S*. Such a structure exists if we are able to define the Poisson bracket between two functions on *S*. Manifolds with such a structure are called *symplectic manifolds*, the expression phase space being employed for symplectic manifolds of standard classical mechanics. As everybody knows, an ordinary phase space is built from a configuration space *B* and the corresponding speed space (one at each point of *B*). In the mathematical jargon, this phase space is the tangent bundle with *B* as a base and the speed space as a fiber. A symplectic manifold is a more general being which looks *locally* like a phase space[5]. This means that, locally, one can define coordinates $q_1, q_2,...,q_n, p_1, p_2,...,p_n$ for a point x such that the Poisson bracket $\{f(x), g(x)\}$ is given by

$$\{f(x), g(x)\} = \sum \sigma^{ij}(x) \, \partial_i f(x) \, \partial_j g(x) \qquad (3,2)$$

where $\sigma^{ij} = - \sigma^{ji}$. The tensor σ is called a symplectic form. It permits to define the *symplectic scalar product* of two tangent vectors[6]

$$(dx, \delta x) = - (\delta x, dx) = \sum \sigma_{ij}(x) \, dx^i \, \delta x^j \qquad (3,3)$$

The simplest example of a non trivial symplectic manifold is provided by an ordinary sphere. If *u* denotes a point of the sphere and *du*, *δu* two tangent vectors at this point,

---

[4]More rigorously, the structure is that of a projective Hilbert space (a state is a ray, that is a unit vector defined up to a phase). The cohomology of the group is such that every unitary projective representation is provided by a unitary action of $\wp$ (or its covering group) on the underlying Hilbert space. The situation is more subtle in the case of the Galilei group.

[5]In particular, it is even dimensional. The corresponding canonical transformations are often called symplectomorphisms.

[6]Remember that if we define on a manifold a non degenerate symmetric form $g_{ij}(x) = g_{ji}(x)$, it provides us with a scalar product of two tangent vectors *dx* and *δx: (dx, δx) = $\sum g_{ij}(x) \, dx^i \, \delta x^j$*. The manifold is a Riemannian manifold.

the symplectic scalar product is given by

$$(du, \delta u) = u \cdot (du \times \delta u) \tag{3,4}$$

Clearly, this is an antisymmetric product; if the radius of the sphere is one, the expression (3,4) is the familiar surface element of the sphere $sin\vartheta \, d\vartheta \, d\varphi = - dz \, d\varphi$ where $z = cos\vartheta$. It follows that φ and z are local canonical coordinates (they are not defined for $z = \pm 1$).

Quantum free elementary system	Classical free elementary system
Poincaré invariance.	Poincaré invariance.
The set of states is a projective Hilbert space.	The set of states is a generalized phase space (a symplectic manifold).
℘ acts unitarily.	℘ acts by symplectomorphisms.
℘ acts irreducibly.	℘ acts transitively.

Table 1

It is clear that the symplectic scalar product (3.4) is invariant under any rotation of the sphere; therefore, the rotations preserve the symplectic structure of the sphere: *they are canonical transformations*. This has an important consequence for the classical momentum associated with the rotation group, namely the angular momentum J. Let us fix the length of the angular momentum. The possible values of this momentum form a sphere of radius $|J|$. In terms of the canonical symplectic coordinates φ, J_z, the components of J are:

$$J_x = \sqrt{|J|^2 - J_z^2} \, cos\varphi$$

$$J_y = \sqrt{|J|^2 - J_z^2} \, sin\varphi \tag{3,5}$$

$$J_z$$

The Poisson bracket $\{\varphi, J_z\}$ being equal to 1, the reader will have no difficulty to check that

$$\{J_x, J_y\} = J_z \, , \quad \{J_y, J_z\} = J_x \, , \quad \{J_z, J_x\} = J_y \tag{3,6}$$

This calculation was made by Kramers to propose a classical model for spin; the sphere being two-dimensional, Kramers declared that spin was corresponding to *one degree of freedom* .[7]

Let us come back to the role of the Poincaré group in the definition of an elementary system. This role is summarized in Table 1.

The problem to be solved is the following one: how to find a symplectic homogeneous space of the Poincaré group such that the group acts in preserving the symplectic structure. A simple theorem due to Kirillov gives all symplectic homogeneous spaces of any Lie group[8]. They are known as *coadjoint orbits because* they are the orbits of the group acting on the dual vector space of the Lie algebra (the orbits in the Lie algebra itself are the adjoint orbits; for *simple* Lie groups, as the rotation group, the adjoint and coadjoint orbits can be identified with the aid of the Killing form). Let us explain concretely, in a simple case (a non simple group) the difference between the Lie algebra and its dual vector space. We consider the Euclidean group of the ordinary space. The physicists denote usually by J_i and P_i the "infinitesimal generators" of this group; the generic element of the Lie algebra can be written $J.\omega + P.a$. As a vector space, the Lie algebra is made of couples (ω, a) and its dual is made of couples (J, P). The Euclidean group being not simple, we must expect a different action of it on the Lie algebra and its dual. Let us perform, for instance, a translation b. We have

$$exp(-iP.b) \ (J.\omega + P.a) \ exp(iP.b) = J.\omega + P.a + (b, \omega, P) \tag{3,7}$$

where the last parenthesis denotes the mixed vector product. This implies that the transformation is given by

$$(\omega, a) \qquad \rightarrow (\omega, a + b \times \omega) \qquad \text{"rotational part" unchanged}$$

but $\tag{3,8}$

$$(J, P) \qquad \rightarrow (J - b \times P, P) \qquad \text{"translational part" unchanged}$$

If we had performed a rotation R followed by a translation b, that is the most general Euclidean transformation, the formulas (3,8) would have the vectors ω, a, J and P rotated by R. Let us concentrate on the coadjoint action. First, the point $(J, P) = (0, 0)$ is invariant; the corresponding orbit is just that point; it is a trivial symplectic manifold of dimension zero. The point $(J, 0)$ with $|J| \neq 0$ is transformed into $(RJ, 0)$; the orbit is a sphere of radius $|J|$. Finally, it is possible to prove that the other orbits are of dimension four: they are defined by the equations

$$P^2 = \text{constant (non zero)} \quad \text{and} \ (J.P)/|P| = \text{constant} \tag{3,9}$$

---

[7]It is interesting to underline that the square root of $|J|^2 - J_z^2$ has its quantum counterpart, the square root of $j(j+1) - m(m+1)$.

[8]See Appendix A for a proof that every coadjoint orbit is symplectic.

One recognizes the kinetic energy and the helicity. We see that the helicity observable is already obtained with the Euclidean subgroup of the Poincaré group.

We arrive now at the coadjoint orbits of the Poincaré group itself. Practically, they can be described in the following way, quite analogous to the one used for the Euclidean group. Instead of (J, P) in a 6-dimensional space, we have a couple $(M_{\mu\nu}, P_\rho)$ in a 10-dimensional space. If we discard the case where the four-vector P_ρ is zero[9], the classification of orbits is obtained with the aid of the Pauli-Lubanski vector:

$$W_\lambda = 1/2 \; \varepsilon_{\lambda\mu\nu\rho} \, M^{\mu\nu} \, P^\rho, \text{ where } \varepsilon \text{ is antisymmetric and } \varepsilon_{0123} = 1 \qquad (3,10)$$

by the following table, where $T, L, S, 0$ denote the type of 4-vectors, namely time-like, light-like, space-like and zero.

P_ρ	W_λ	dimension of the orbit	interpretation
T	S	eight	spinning massive particles
T	0	six	spinless massive particles
L	L	six	massless particles with helicity
L	0	six	massless spinless particles (unphysical?)
L	S	eight	unphysical
S	T	eight	"
S	L	eight	"
S	S	eight	spinning tachyons
S	0	six	spinless tachyons

Table 2

Let us note that the $(T)(0)$-orbit gives back to the usual six-dimensional phase space of classical point particles[10]; the $(T)(S)$-orbit is a sphere bundle on $\mathbb{R}^6$, the sphere being identified with the Kramers spin sphere. The phase space for a spinning point particle is obtained from the ordinary six-dimensional phase space in associating a Kramers'sphere

---

[9]When P_ρ is zero, we are left with the classification of the coadjoint orbits of the Lorentz group. Since it is a simple group, we could classify the adjoint orbits as well. This is equivalent to classifying the (homogeneous) electromagnetic fields (E, B). Because of the two invariants $E^2 - B^2$ and $E.B$, the general orbit is four-dimensional. There is a single orbit of dimension two (when both invariants are zero).

[10]In non relativistic mechanics, the ordinary phase space is the homogeneous space G/H where G is the Galilei group and H the group generated by time translations and rotations.

with each point. We must recall that the spin and the helicity can take any value in this classical model of elementary particles.It is also important to emphasize that, contrary to the classical approach of relativistic classical mechanics, space-time is not an ingredient of the theory. The question of rediscovering space-time and worldlines from the coadjoint orbits will be discussed later on. The advantage of this approach is to give more importance to the momenta. The fact that they obey laws of conservation and additivity makes this theory closer to experimental aspects. Finally, I must say a few words about the quantization problem itself. It is clear that quantizing means here finding a unirrep of the Poincaré group. The Kostant-Souriau quantization procedure itself can be used to build explicitly a representation of a given Lie group from an arbitrary coadjoint orbit.[11] This can be used to solve a quantum problem when we know a dynamical symmetry group of it.

Bibliography

[Are1]   R. Arens, Comm. Math. Phys. *21*, 125 (1971).
[Are2]   R. Arens, Comm. Math. Phys. *21*, 139 (1971).
[Are3]   R. Arens, J. Math. Phys. *12*, 2415 (1971).
[Bac1]   H. Bacry, *Classical Hamiltonian Formalism for Spin* (unpublished, 1966).
[Bac2]   H. Bacry, Comm. Math. Phys. *5*, 97 (1967).
[Kos]   B. Kostant, Lecture Notes in Mathematics, 170 (Springer-Verlag, Berlin, 1970).
[Dir]   P.A.M. Dirac, *Directions in Physics* (John Wiley and Sons, New York).
[Kra]   H.A.Kramers, *Quantum Mechanics*, p.235 (North Holland, Amsterdam, 1958).
[Lub]   J.K. Lubanski, Physica *9*, 310 (1942).
[S.W]   D.J. Simms and N.M.J. Woodhouse, *Lectures on Geometric Quantization*(Springer-Verlag, Berlin, 1976).
[Sou1]   J.-M. Souriau, Cptes Rend. Ac. Sci. *263*, B1191 (1966).
[Sou2]   J.-M. Souriau, *Structure des systèmes dynamiques* (Dunod, Paris,1970).
[Wig1]   E.P. Wigner, Ann. Math. *40*, 149 (1939).

---

[11]From my point of view, when I suggested this classical model for elementary particles, the problem of quantization was solved *ipso facto*, quantizing becoming equivalent to finding a unitary representation of the Poincaré group. But quantization is not only a *mapping* associating a quantum object to its classical counterpart; it also means a *procedure* which permits to associate a Hilbert space with a symplectic manifold in such a way that the symplectomorphisms (i.e. the canonical transformations) are represented unitarily. The Kostant-Souriau quantization is such a procedure. It was discovered independently by these two authors, the first having in mind the building of representations of Lie groups, the second the realization of the Dirac programme of replacement of Poisson brackets by commutators. We give in Appendix *B*, as an example, the quantization of the sphere *à la Souriau*.

CHAPTER 4

LOCALIZABILITY. THE PHOTON SCANDAL.
QUANTIZATION HELPLESS!

The photon is not localizable! It is not exagerate to say that almost every physicist knows this fact but does not care. A position operator is not an important object. The important operators in quantum physics are the energy, the linear and angular momenta. The spectroscopist, whatever is his field (particle, nuclear or atomic), is not concerned with position! The position operator is only for students and, more precisely, only for beginners in quantum mechanics... and for people interested in the sex of the angels, this kind of people you find among mathematical physicists, even among the brightest ones as Schrödinger or Wigner...

The photon is not localizable!

Why is it so? What does it mean? It is the purpose of this chapter to examine in some detail this situation. First I will give a very simple argument to show why the photon cannot be localized. If the photon was localizable, the squared modulus of its associated wave function, namely the intensity of the Maxwell field would provide us with a probability density of finding the photon in some place. That it is impossible is a direct consequence of dimensional analysis: no quantity of the kind

$$h^m c^n e^p A^2$$

where A is the vector potential and h, c, e are the usual fundamental constants, can be given the dimensions of the inverse of a volume. The replacing of A^2 by $E^2 + B^2$ would not solve the difficulty.[1]

---

[1] I tried vainly to find such a simple argument in literature. See [B.L.P], p.12 for a longer argument based on covariance (no reference to the Newton-Wigner's notion of localizability in this textbook). It is amusing to note that there are textbooks stating that $E^2 + B^2$ is proportional to the probability density of finding a photon in some place, but they are very prudent not to give the factor of proportionality. Dimensional analysis would oblige us to use the gravitational constant in this factor. There is another fact relating electrodynamics and dimensional analysis: there is no scale in Maxwell equations, a fact which is related with the conformal invariance. I am grateful to L.C. Biedenharn who drew my attention to the Casimir paradox:" *Let j(x,ω) and ρ(x,ω) be charge-current density distributions, varying with time as exp(-iωt), and vanishing outside a sphere of radius R. Then it is always possible to find another charge-current density (j₁,ρ₁) vanishing outside a radius R₁<R such that the radiation field outside R is identical to that produced by the original sources*" [B.L1]. We find in this reference a picturesque way of

There exists another argument I have found in the literature [Kra] but which is not convincing. I will say why. Suppose that we want to describe the photon states by the transverse potential A (k)

$$k. A (k) = 0$$

It is clear that the operator $i \frac{\partial}{\partial k}$ does not preserve the transversality condition since

$$k. i \frac{\partial}{\partial k} A = i \frac{\partial}{\partial k} (k.A) - iA \neq 0$$

This argument only proves that the operator $i \frac{\partial}{\partial k}$ cannot be interpreted as a position operator but nothing can be concluded about the non localizability of the photon. However, we will see that it is the transverse character of the wave function (related with the spin of the photon) which is responsible of the non localizability property. From the Newton-Wigner study of localized states, it is clear that if the photon was spinless it would have localized states!

Before recalling the postulates of Newton and Wigner and the ones of Wightman, let us discuss the meaning of localizability and its relationship with the position operator. Unfortunately, very often, the two notions are strongly associated and it is said, for instance, that the photon is not localizable *because* it has no position operator. In fact, to have a localized state means that, given a domain S in the regular space, there is a state for which we can say that the probability for the particle to be in S equals one. From Newton and Wigner's work, if a particle has localized states, it follows that one can define a set of three *commuting* operatorswhich can be interpreted as the components of the position operator. Therefore, if a particle has no localized states, we have the following alternative: either it is impossible to measure any coordinate, that is there is no position operator, or the position operator has three non commuting components. We will see the advantages we get in choosing the second part of the alternative.

Let us give a brief description of the Newton-Wigner postulates. The authors supposed that the states which represent a particle[2] in a state localized at $x = y = z = 0$ obey the following conditions:
a) they form a linear set S_0.
b) S_0 is invariant under rotations around the origin and reflections both of the spatial and of the time coordinates,
c) if ψ is a state of S_0, a space translation of ψ shall make it orthogonal to all states of S_0,
d) some regularity conditions under boosts.

---

stating this theorem and due to Casimir himself:*"Suppose an elephant in a spherical cage is illuminated only by coherent light sources inside the cage. Then the spectators outside the cage cannot be sure that there really is an elephant: the cage might be empty but for a peculiar charge-current density at the center of the cage."*
[2]or, more generally, an elementary system (as in its paper of 1939, Wigner is insisting on this notion).

All these conditions seem very natural; the dimension of the space S_0 is left arbitrary to include spin variables. Obviously, localized states are not elements of the Hilbert space and orthogonality is understood in the generalized sense involving the delta function. Wightman postulates, although inspired by Newton's and Wigner's ones, were more rigorous and concerned Hilbert spaces carrying representations (irreducible or not) of the Poincaré or Galilei group. This is in contradistinction of the Newton-Wigner approach were spinning particles were described by Bargmann-Wigner equations. The Wightman postulates were the following ones. If S_i denotes any Borel set in the ordinary space (denoted by R^3) and $E(S_i)$ the projection on the Hilbert subspace of all states where the system is localized in S_i), we must have:[3]

i) $E(S_1 \cap S_2) = E(S_1) E(S_2)$,

ii) $E(S_1 \cup S_2) = E(S_1) + E(S_2) - E(S_1 \cap S_2)$,

iii) $E(S_1 \cup S_2 \cup S_3 \ldots) = E(S_1) + E(S_2) + E(S_3) + \ldots$ for disjoint Borel sets,

iv) $E(\mathbb{R}^3) = I$ (the unit operator),

v) $E(RS_i + a) = U(a, R) E(S) U(a, R)^{-1}$, where (a, R) denotes a Euclidean transformation in the unitary representation of the kinematical group,

vi) regularity conditions for boosts,

vii) invariance with respect to time reversal.

Let us give a résumé of the main conclusions of Newton and Wigner for relativistic particles:

a) localized states exist for all spinless particles (including the massless ones),

b) they also exist for spinning massive particles with the following "disease": if a state is localized for one observer, it is no longer localized for another one (localized states are not transformed into localized states under a Poincaré boost),

c) they do not exist for massless particles with helicity (except for a four component neutrino). In particular, there is no localized state for a photon nor for a two component neutrino.

Let us make some comments about these results.

a) A localized state for a spinless particle of mass m is not described by a delta function, but rather by

$$\psi (r) = (m/r)^{5/4} H_{5/4}(i\, mcr)$$

where H denotes a Hankel function. The corresponding position operator (the so-called Newton-Wigner operator for a scalar particle) is given by

$$q = i\frac{\partial}{\partial p} - i \frac{p}{2p_0^2}$$

The departure from the usual Schrödinger operator is only due to the fact that q must be

---

[3]The system of projections form a *system of imprimitivity* in the language of Mackey.

Hermitian with respect to the invariant measure $\dfrac{d^3 p}{p_0}$ of relativity, instead of the measure $d^3 p$. That explains why the Hankel function has replaced the traditional Dirac delta function. About this formula, Wigner writes, after having underlined that it is the Fourier transform of $\psi(p) = \sqrt{p_0}$: "*$\psi(p)$ should be orthogonal not only to the wave functions that arise from it by purely spatial displacement, i.e., to $\psi_x(p) = \sqrt{p_0}\, exp(ip.x)$ ... but also to those that result from it by an additional time displacement by t, $\psi_{x,t}(p) = \sqrt{p_0}\, exp(i(p.x - p_0 t))$, as long as the space-time vector is within the light cone, i.e. as long as ct<x. This is not the case and this shows that at least particles of spin zero cannot be truly localized. And the situation is pretty much the same for higher spins. This is the other reason I believe our present idealized space-time concept will undergo modification*" [Wig3]. This is a very pessimistic (recent) conclusion for the work made in collaboration with Newton.

Obviously, the delta function is instinctively associated with a straight worldline in Minkowski space-time and, already in classical relativity, physicists encountered difficulties in trying to associate a given worldline to a particle with extension, that is to say to define in a canonical way a worldline in a "worldtube". There is no "center of mass motion" on which all observers would agree. In other words, there is no way of assigning a sharp worldline with a given particle with internal structure.

b) For spinning particles, the situation is worse in classical relativity because the center of mass does not coincide, for a given observer, with the center of rotation! [Pry]. We will not enter into details about the expression of the N.W. operator for spinning particles.

c) Conclusion *c* is difficult to accept, at least for three reasons. First, there are many experimental ways of detecting photons in some place and we must be able to define a position operator on the Hilbert space of the photon states. Second, if we want to have all kinds of particles on the same footing, we must have a position operator for all of them. Third, Einstein built Minkowski space-time with the aid of light signals; it follows that there must exist a way of deriving space-time structure from the Hilbert space of the photon states. It is remarkable, however, that Einstein's light signals were made of *scalar waves* and that the non localizability is precisely due to the polarization of light! We are faced with a historical paradox: *in 1905, Einstein building Minkowski space-time with the aid of light signals and, simultaneously, endowing light with a structure forbidding him to build it!* We will analyze this situation later on.

It is interesting to mention that all the difficulties encountered by Newton and Wigner are also present in the underlying coadjoint orbit model of classical particles. Without entering details, the situation is as follows.
a) For spinless particles (even massless), it is possible to associate a given straight worldline in the Minkowski space-time with each point of the coadjoint orbit.

b) For a spinning massive particle, each given Galilean observer can associate with a given classical state a well defined worldline. Unfortunately, all these worldlines do not coincide and lie in a tube which has an extension of the order of the Compton wave length.

c) For a massless particle with non zero helicity, the situation is really a scandal because the localization becomes worse: the wordlines associated with two Galilean observers are at a distance of the order of the wave length, but the wave length is, as everybody learnt from the Doppler effect, as large as we want! The set of all worldlines has an infinite extension in space-time![4] In other words, there is no way of deriving the Maxwell field (i.e. the first quantization theory of the photon) from a classical model; *there is no satisfactory quantization procedure for the one photon states.* A very simple calculation about this fact is given in Appendix D.

History of physics is responsible of the fact that the first quantization of the photon is usually ignored in textbooks and the solutions of Maxwell equations are very rarely presented as describing the Hilbert space of the photon states.[5] Photons are rather introduced as excitations of the Maxwell field or as modes in a cavity. More surprisingly, the photons are very often supposed to be *always* in an eigenstate of the energy-momentum[6], a fact which is in contradiction not only with the Hilbert space structure of the set of a one particle quantum states (the superposition principle) but also with experiments since photons have a natural width in their spectrum. In many textbooks, after having defined photons as eigenstates of the energy-momentum, the authors are referring to photons in a given angular momentum state, as if a photon could be simultaneously in an eigenstate of both angular and linear momenta (two non commuting observables)! [7]

Another surprising fact concerns the *speed* of the photon. The speed is a classical concept, the photon is not. How to associate two such concepts? In ordinary quantum mechanics, the speed of a particle is obtained by using the commutator of the position operator with the Hamiltonian. Since the photon is said not to have any position operator, this is not a way to define the speed of the photon. There is no logical way to arrive at the ratio *p/H,* as it is the case for a massive particle[8]. The only definition we are left with for the speed in quantum mechanics was given by de Broglie who interpreted it as the

---

[4]I was aware of the difficulties *b* and *c* as soon I proposed the Poincaré group model of elementary particles. My feeling at that time was that in order to solve the difficulty, we had to reject the Poincaré group itself. To-day, I am convinced that we have to keep it, but to only reject Minkowski's space-time, as it will be discussed later on.

[5]One of the nicest exceptions is provided by reference [Bia], a textbook which has the advantage of being rigorous and giving all quantization procedures of the Maxwell field.

[6]Let us quote, for instance, Pais: "*A photon is a state of the electromagnetic field with the following properties 1. It has a definite frequency* v *and a definite wave vector k. 2. Its energy E, E = h v and its momentum p, p = h k satisfy the dispersion law E = c |p| characteristic of a particle of zero rest mass. 3. It has spin one...The single particle states are uniquely specified by these three properties*"[Pai1, p. 407]. I quoted Pais because he gives here a concise idea of what is in mind of almost all physicists.

[7]Surprisingly, this mistake can be found in textbooks written by quite well known physicists.

[8]We could explain the attitude of some physicists about the lack of localizability of the photon in the following way. If a photon is always in a given energy-momentum state, it corresponds to a plane wave state which fills up the whole space homogeneously; it follows that the photon is everywhere...

group velocity. It is well known that an *arbitrary* wave packet of electromagnetic waves has a group velocity which is *less than c*. Therefore, according to what we have learned in special relativity, there exists *always* an observer for which *the photon is at rest.*.[9] This is not paradoxical. The superposition principle tells us that a photon can be in a stationary wave state, say $e^{ikz}+e^{-ikz} = 2\cos kz$. What would be the speed of such a photon? If it was c, what would be the direction of this speed? Towards the positive or the negative z axis? It is clear that the only correct answer is that such a photon has speed zero[10]. All these remarks justify amply the last part of Einstein's sentence given at the end of Table 4 of Chapter 1.[11]

In order to illustrate all the difficulties encountered by the photon concept, let us quote and comment parts of an interesting text of M. Sachs, in [Hoo]. It starts as follows: *"A very old, yet unresolved problem in physics concerns the basic nature of light... Still, logical dichotomy and mathematical inconsistency remain in the usual answers to the question: What, precisely, is light?"* A few pages later, he is underlying conceptual difficulties: *"But a single photon, which, by definition, has a precise energy, is described mathematically in terms of a plane wave - a function that has an equally weighted value at all points in space at any given time. With this description, then, one would have to say that the single photon is everywhere, rather than somewhere - although it can be annihilated somewhere by looking for it at that particular place! Along with this spatial description of the single photon, it is specified to be continually travelling at the speed of light. To the (perhaps naive) inquirer, the logical difficulty appears in trying to answer the question: if the photon is everywhere at the same time, and is travelling continually on its own at the speed of light, where is it going to?"*

This quotation has the interest to underline tha danger of the two standard statements of textbooks: *a*)the photon travels at the speed of light; *b*) it has a given energy. In rejecting these two properties (which contradict the postulates of quantum theory), the difficulties described by Sachs disappeared.[12]

---

[9] Heitler [Hei, p.16] arrives to a conclusion closed to mine, but stated prudently:*"If the[electromagnetic]field differs from zero only with a certain given volume V, and if no charges are present inside the volume,... the momentum G and energy U of the field form a 4-vector G_μ which behaves as regards its transformation properties like the energy and momentum of a particle... A spherical light wave emitted from a point source... has a momentum zero but a finite energy."*

[10] a fact which is not accepted even in excellent textbooks: *"There is no inertial frame in which the photon is at rest."* [Wic,Section 4.7]. This refusal is due to the frequent confusion between a frame (in the ordinary space, which is an affine space) and a basis (in momentum space, which is a vector space). See Chapter 8, Table 1 of the present book. Unfortunately, the expression "rest frame of a particle" is very common but not satisfactory in quantum physics: "frame" implies a position (the origin of the frame) and "rest" implies a zero momentum. But momentum and position cannot be defined simultaneously. In fact, what physicists have in mind when they speak of a "rest frame" is a rest basis in momentum space.

[11] See [Str] for difficulties encountered in textbooks about the photon concept. (note that the author of this article misunderstood the concept of localizability of particles). It would be tedious to quote all contradictions which can be found in the main textbooks.

[12] In the same text, Sachs also says:*"[The photons] cannot be slowed down or speeded up because they have no inertial mass; They can only be stopped, by annihilating them... they have the peculiar un-particle-like feature of being without inertia since no external force can make them change their speed by arbitrary amounts."* Such a remark is really strange. In general relativity, we know that the source of the

There is one result of the Newton and Wigner paper which is of interest and which ll be discussed later on. It concerns the Newton-Wigner position operator for a Dirac rticle. It has a complicated expression we will not reproduce here. It satisfies the lations:

$$[q_i, q_j] = 0$$

$$\frac{dq}{dt} = i\,[H,q] = i\,[c\alpha.p + \beta mc^2, q] = H^{-1}p$$

tead of

$$\frac{dx}{dt} = c\alpha$$

iich is unacceptable since the eigenvalues of any component of the r.h.s. are $\pm c$.

Bibliography.

.L1]   L.C. Biedenharn and J.D. Louck, *Angular Momentum in Quantum Physics, Theory and Application* (Addison-Wesley, Reading, Massachussets, 1981).

.L.P]   V.B. Berestetskii, E.M. Lifshitz and L.P. Pitaevskii, *Relativistic Quantum Theory* (Pergamon Press, Oxford, 1971).

ia]   I. Bialynicki-Birula and Z. Bialynicka-Birula, *Quantum Electrodynamics* (Pergamon Press, Warsaw, 1975).

dd]   A.S. Eddington, *Fundamental Theory* (Cambridge University Press, London, 1946).

ok]   A.D. Fokker, *Relativitatstheory* (Groningen, Noordhoff, 1929).

ei]   W. Heitler, *Quantum Theory of Radiation* (Oxford University Press, Oxford, 1954).

oo]   C.A.Hooker,Ed.,*Contemporary Research in the Foundations and Philosophy of Quantum Theory* (Reidel, Dordrecht-Holland, 1973).

.W]   T.D. Newton and E.P. Wigner, Rev. Mod. Phys. *21*, 400 (1949).

ii]   A. Pais, *Subtle is the Lord* (Clarendon Press, Oxford, 1982).

i]   W.C. Price and S.S. Chissick, Ed., *The Uncertainty Principle and Foundations of Quantum Mechanics* (Wiley,London,1977).

ry]   M.H.L. Pryce, Proc. Roy. Soc. A*195*, 62 (1948).

r]   J. Strnad, Am. Journ. Phys. *54*, 650 (1986).

ic]   E.H. Wichmann, *Quantum Physics* (Mac Graw Hill, New York, 1968).

ig]   A.S. Wightman, Rev. Mod. Phys. *34*, 845 (1962).

ig2]   E.P. Wigner, Int. Jour. Theor. Phys. *25*, 467 (1986).

vitational field is the energy-momentum, not the mass; it follows that inertia is also described by the rgy-momentum. If we say that to be *at rest* means that the momentum is zero, the photon cannot be est, but its speed can have any value less than or equal to c.

CHAPTER 5

A POSITION OPERATOR FOR THE PHOTON.
GIVING UP THE COMPLEMENTARITY PRINCIPLE.

Some years ago, in teaching symmetry at the physics department of the Technion in Haifa, I decided to illustrate the classical Noether theorem in treating the problem of an electrically charged particle in the field of a Dirac magnetic monopole; since the magnetic field

$$B = gr^{-3}r \qquad (5,1)$$

has the same magnitude in all directions, we have a spherical symmetry which implies, from Noether's theorem, the conservation of a vector which is naturally called the angular momentum J. This vector can be obtained in a gauge invariant way[1], without specifying the vector potential. Its expression is well known; it is

$$J = r \times (p - sA) - s\frac{r}{r} \qquad (5,2)$$

where we set, for convenience,

$$B = g \, curl \, A, \, s = eg \qquad (5,3)$$

One easily verifies that the components of J satisfy the Poisson brackets relations of the angular momentum. To get the expression of the angular momentum in quantum mechanics, we only have to replace r and p by their usual corresponding operators. *It follows* that the product $s = eg$ is necessarily quantized: it is a multiple of $1/2$. That it is just a consequence is not trivial because it does not follow from the commutation relations alone but, rather, from the operators r and p themselves.

Remembering a beautiful work by Lipkin, Weisberger and Peshkin [L.W.P] who used the commutation relations of J_i and the r_i and the properties of the Euclidean group generated by these operators, I tried to improve their proof in choosing a larger group, in order to shorten the argument. I was pleased to see that the operators J_i, r_i, $r = |r|$ and K_i, where[2]

---

[1] See Appendix C.
[2] J_i and K_i are the infinitesimal generators of the Lorentz group. If we denote by $M_{\mu\nu}$ these generators, the J_i generate the rotations ("magnetic part" of $M_{\mu\nu}$) and the K_i the boosts ("electric part" of $M_{\mu\nu}$ with a change of sign).

$$K = \frac{1}{2} [r(p - sA) + (p - sA)r] \tag{5,4}$$

an the Lie algebra of the Poincaré group, with an odd interpretation since (r, r_i) plays
: role of the energy-momentum. The "mass" (the square root of $r^2 - r^2$) is obviously
:o and the "helicity operator" is just a number given by

$$\frac{J.r}{r} = s \tag{5,5}$$

ace we know, from Wigner's work on the Poincaré group representations, that the
licity is quantized, s is certainly an integer or half an integer.[3]

Let us now perform the following canonical transformation of the representation of
: Poincaré group we have just arrived at:

$$r \rightarrow - p$$
$$\tag{5,6}$$
$$p \rightarrow + r$$

ie expression $p - s A(r)$ becomes $r - s A(-p)$ and the generators now read

$$J = (r - s A (-p)) \times p + s\frac{p}{p} \tag{5,7}$$

$$K = \frac{1}{2} [p(r - sA(-p)) + (r - sA(-p))p] \tag{5,8}$$

d let us now interpret the translation generators as the energy and the momentum of a
rticle which is massless with helicity s. In examining the expression of J, we are
npted to say that $R = r - sA(-p)$ plays the role of the position operator and it is natural
investigate the validity of such an interpretation since we know that such particles have
> Newton-Wigner position operator.

First, we note that, according to Eqs (5,1) and (5,3), when s tends towards zero, R
:comes just r, the usual position operator for a spinless particle. It follows that the
fference $R - r$ is a small correction due to the spinning property. Now, from what we
ve seen in the last chapter, if R has the interpretation of the position operator, its
mponents cannot commute. In fact, we have

$$[R_i, R_j] = - is \, \varepsilon_{ijk}\frac{p_k}{p^3} \tag{5,9}$$

he product eg could be interpreted as the "radial component" of the angular momentum. Heuristically,
: could say that its spectrum is discrete as it is the case for any component. However, this is not a
>of: the radial component is in fact a *scalar operator* under rotations, that is it commutes with the three
mponents of J; therefore, the commutation relations cannot give any information on its spectrum.

We know the impossibility for a massles particle to have $p = 0$. It follows that there is no state for which the R_i 's are commuting and, consequently, such a particle has no localized state, a result which is in agreement with the Newton-Wigner result[4]. If we are interested in the photon, which has two helicity states, we only have to replace s by the helicity operator (eigenvalues ± 1).

Let us show that this operator is a reasonable candidate as a position operator for spinning massless particles.

a) There exists already a situation where non commuting components occur for a position operator: I am referring to the coordinates of the center of the circular trajectory of an electron in a homogeneous magnetic field [L.L, Chapter XVI]. In my knowledge, no objection has been made about such a situation. This can be considered as a good reason to accept (or at least not to reject) the same property for the position of a particle. As we saw in our problem, the non commutativity comes from the helicity of the particle; this is a small effect as it can be seen from Equation (5,9) in introducing explicitly the reduced Planck constant h.

$$[R_i , R_j] = - i \, \hbar s \, \varepsilon_{ijk} \frac{p_k}{p^3} \qquad (5,10)$$

Let us consider the photon case; s has $\pm \hbar$ as eigenvalues; it follows that the modulus of the right hand side of Eq.(5.10) is at most equal to $\lambda^2 / 4\pi^2$, where λ is the wave-length h/p. This has a physical consequence which is quite acceptable: localizability, that is the possibility of measuring simultaneously the three coordinates of the position, is increasing together with the frequency of the photon. In other words, the particle character of the electromagnetic field is sharper and sharper when the frequency of the wave is increasing. Conversely, the notion of a trajectory becomes more and more fuzzy when the wave-length is increasing. Such a conclusion is different from the *complementary principle* because the photon is always present, whatever is its energy, but the *classical character of a point particle* , which is an approximate description, is a good or a bad description, depending on the energy of the photon. The goodness of the "classical trajectory" approximation can vary continuously.

b) Let us consider a photon which is *almost* in an eigenstate of the component p_3 of the momentum. The uncertainty relation between R_1 and R_2 reads

$$\Delta R_1 \, \Delta R_2 \geq \hbar \, | \langle \frac{p_3}{p^3} \rangle | \sim \lambda^2 \qquad (5,11)$$

This means that when we know that a photon is running in a given direction, its transverse position cannot be defined exactly.[5]

---

[4]Obviously, here, in contradistinction with the paper of Newton and Wigner, the non localizability does not imply the non existence of a position operator.
[5]It must be underlined, however, that R and R_2, separately, commute with p_3.

It turns out that Eq.(5,11) has an important consequence about the electron slit experiment. The reader has probably kept in mind the criticism developed in the previous chapter on this subject. Any photon scattered by an electron will have a fuzzy localization, translating itself a fuzzy localization of the electron. This fuzziness increases with the wave-length. Owing to Eq.(5,11), we no longer need the classical wave theory to explain the reappearing of the interference pattern when the wave-length of the light source is increasing. The photon language is sufficient, whatever is its energy.

Before investigating in a deeper way the proposal of a new position operator, it is worthwhile to say a few words about the duality between the spinning massless particle and the Dirac magnetic monopole. Here, as in [Bac3], this duality was presented with the aid of the Poincaré group. It was also discussed by Schwinger in the introduction of [Schw2] and in a more recent article by Barut and Bracken [B.B], but both without any reference to a position operator.

When I wrote my first article on that subject [Bac3], I was probably a little shrinking. I was convinced that I had a *nice* proposal, but I did not know what to do with it. It is only after exciting discussions with Connes that I tried really to understand what would be the consequences of such a new operator for the photon. I was led to state the two following facts:
    i) the needness of changing the Newton-Wigner-Wightman (NWW) axioms for *all* particles.
    ii) the new position operator for the photon was the analogue, for the Maxwell equations, of the position operator proposed by Schrödinger for the Dirac equation.

Then I decided to publish these new conceptual observations in a new article [Bac4]. As soon as it was sent to the Editor, I became aware of two new things:
    i) The above photon position operator had already been introduced in the excellent book by Bialynicki-Birula and Bialynicka-Birula [Bia], but without referring to it as a position operator. They denoted it by $i D$.
    ii) The same position operator has been discovered long time before me by Jadczyk and Jancewicz[J.J][6]. More precisely, these authors defined a position operator for any spin one particle. Their operator is a particular case of the general one proposed in [Bac4] and which will be discussed in the next chapter.

It is interesting to underline that the Bials[7] introduced their operator D in their chapter on the *classical* electromagnetic field; it plays an interesting role in the Fourier analysis of the field. Let us examine why; it will help to understand the meaning of our $A(-p)$.

For this purpose we need a lemma.
Lemma: If n denotes a unit (real) vector, every complex vector $y = y_1 + i y_2$ satisfying

---

[6]I am grateful to D. Kastler for having mentioned this work to me.
[7]I hope that these physicists will forgive me for having unified their name in such a short way.

$$\boldsymbol{n} \times \boldsymbol{y} = -i\,\boldsymbol{y} \qquad \text{and } \boldsymbol{y^*}.\boldsymbol{y} = y_1{}^2 + y_2{}^2 = 1 \tag{5,12}$$

is obtained from a direct orthonormal basis $(\boldsymbol{y_1}, \boldsymbol{y_2}, \boldsymbol{n})$.

The proof is easy and is left to the reader. We can also see readily that two solutions only differ from a phase factor; therefore, this phase factor can be interpreted as a rotation in the plane orthogonal to $\boldsymbol{n}$.

This lemma has the following corollary.
Proposition: Two self-dual tensors (i.e. antisymmetric tensors satisfying $a_{\tilde{\mu}v} = -ia_{\mu v}$) which verify the condition

$$p^{\mu} a_{\mu v} = 0 \tag{5,13}$$

where $p_{\mu} = (p, \boldsymbol{p})$ is a (future[8]) light-like vector, only differ from a phase factor.

Here also the proof is simple: if e denotes the electric part of $a_{\mu v}$, the condition of self-duality implies that its magnetic part is $i\,e$; moreover, Eq. (5,13) implies

$$\boldsymbol{n} \times \boldsymbol{e} = -i\,\boldsymbol{e} \quad \text{with } \boldsymbol{n} = \frac{\boldsymbol{p}}{p} \tag{5,14}$$

and the proposition follows from the lemma.

Let $f_{\mu v}(x)$ be the (real) classical Maxwell field satisfying the two equations

$$\partial^{\mu} f_{\mu v} = 0, \quad \partial^{\mu} f_{\tilde{\mu}v} = 0 \tag{5,15}$$

All the information contained in f is also contained in the self-dual field F defined by

$$F = f + i\tilde{f} \tag{5,16}$$

Let us write the Fourier decomposition of F.

$$F_{\mu v}(x) = \int \frac{d^3p}{p} [a_{\mu v}^+(p)e^{-ip.x} + a_{\mu v}^-(p)e^{ip.x}] \tag{5,17}$$

The upscripts + and - correspond to the future and past half cone, respectively. The two tensors a^+ and a^- are both self-dual and satisfy both Eq.(5,13); therefore our proposition applies and we can write

$$a_{\mu v}^+(p) = a_{\mu v}(p)\, f_+(p)$$

$$a_{\mu v}^-(p) = a_{\mu v}(p)\, f_-^*(p) \tag{5,18}$$

---

[8]The time component p_o is positive.

ere $a_{\mu\nu}$ is supposed to be normalized as follows

$$a^{\mu\nu}(p)a^*_{\nu\lambda}(p) = p^\mu p_\lambda \tag{5,19}$$

d is now uniquely defined *up to a phase factor.*

Let us denote by $e(p)$ the electric part of this tensor and let us examine how this ctor varies with p. We may write

$$i\frac{\partial e_j(p)}{\partial p_i} = \alpha_i(p)\, e_j + \beta_i(p)\, e_j^* + \gamma_i(p)\, p_j. \tag{5,20}$$

' using the orthogonality of the vectors e, e^* and p, we obtain

$$\frac{\partial e_j(p)}{\partial p_i} = -i\,\alpha_i(p)e_j(p) - p^{-2}\,e_i(p)p_j. \tag{5,21}$$

we decide to change e by a phase factor $e^{-i\phi(p)}$, the vector field $\alpha(p)$ transforms as lows

$$\alpha(p) \rightarrow \alpha(p) + \nabla\phi(p) \tag{5,22}$$

d such a change induces the following transformations on the functions f_+ and f_-:

$$f_+(p) \rightarrow e^{-i\phi(p)}f_+(p)$$

$$f_-(p) \rightarrow e^{i\phi(p)}\, f_-(p) \tag{5,23}$$

e are led to introduce the covariant derivatives[9] D_i of the functions $f_\pm$:

$$D_j f_s(p) = \frac{\partial}{\partial p_j}f_s(p) - i\,s\alpha_j(p)f_s(p) \tag{5,24}$$

is has the advantage that a change of phase induces on the functions $f_\pm$ the nsformations

$$D_j f_s(p) \rightarrow e^{is\phi(p)}f_s(p) \tag{5,25}$$

such a way that expressions like $f^*D_j f$ or $(D_i f)(D_j f)$ are phase independent. If we mpute the commutator of the derivatives D_i, we obtain

---

ials mention, unfortunately without reference, a work done in 1973 by Staruszkiewicz on the rpretation of these covariant derivatives (p.131).

$$[D_i , D_j] = i\, s\varepsilon_{ijk}\frac{p_k}{p^3} \tag{5,26}$$

that is the iD_i's obey the same commutation relations as our R_i's. In fact, these operators are identical as it can be seen from Eq.(5,24). Bias'vector function $\alpha(p)$ is nothing else than our "vector potential" $-A(-p)$ of Eq.(5,8). This can also be seen in the following way: one can compute the components of the energy-momentum tensor in terms of the field $a_{\mu\nu}(p)$ and the functions $f_{\pm}(p)$ and find from them the generators of the Poincaré group. Bials obtain

$$H = \Sigma\!\int\!\frac{d^3p}{p}\, p f_s^*(p) f_s(p)$$

$$P = \Sigma\!\int\!\frac{d^3p}{p}\, p f_s^*(p) f_s(p)$$

$$\tag{5,27}$$

$$J = \Sigma\!\int\!\frac{d^3p}{p}[f_s^*(p)(i\, D{\times}p) f_s(p) + s\frac{p}{p} f_s^*(p)\, f_s(p)]$$

$$K = \Sigma\!\int\!\frac{d^3p}{p}[f_s^*(p)\frac{i}{2} D f_s(p) + (\frac{i}{2} D f_s(p))^* f_s(p)]$$

which prove our statement.[10]

Eqs (5,27) have an immediate quantum interpretation in QED in replacing f and f^* by the corresponding annihilation and creation operators. In this framework Eqs (5,23) describe the well known fact that the creation and annihilation operators are defined up to a phase. It would be interesting to examine more carefully the problem of the gauge choice in the magnetic monopole vector potential and the role of the Dirac or Schwinger string.

The Bials operator, as my operator R are acting on a two-component object. It is natural to look for an expression of the position operator corresponding to the six-dimensional Maxwell field. There are infinitely many such expressions (all equivalent) because there are many operators acting on a complex vector space-time field $F = B - iE$ which are equivalent once they are restricted to solutions of the Maxwell equations. One of the possible expressions is [Bac4]:

$$R = i\frac{\partial}{dp} + \frac{p{\times}S}{p^2} \tag{5,28}$$

---

[10]Bials go even further since they provide us with the generators of the conformal group [Bia, p.138]. They also study the problem of parity and time reversal.

where

$$S_x = \begin{pmatrix} 0 & 0 & 0 \\ 0 & 0 & -i \\ 0 & i & 0 \end{pmatrix}, \ S_y = \begin{pmatrix} 0 & 0 & i \\ 0 & 0 & 0 \\ -i & 0 & 0 \end{pmatrix}, \ S_z = \begin{pmatrix} 0 & -i & 0 \\ i & 0 & 0 \\ 0 & 0 & 0 \end{pmatrix} \qquad (5,29)$$

From Eq.(5,28), the commutation relations between the R components read

$$[R_x, R_y] = -i\frac{S.p}{p^4}p_z \ \text{(cycl.)} \qquad (5,30)$$

Obviously, the transversality of the field ($p.F = 0$) "kills" the zero eigenvalue of the operator $S.p$ (no "longitudinal" photons). Therefore we are left with

$$[R_x, R_y] = \mp i\frac{p_z}{p^3} \ \text{(cycl.)} \qquad (5,31)$$

which is equivalent to Eq.(5,9).

To conclude, we give here the expression of the field F describing a photon in an eigenstate of R_z with eigenvalue a.

$$F_x(p) = \frac{\pm ipp_y - p_xp_z}{p(p_x^2 + p_y^2)} G(p_x, p_y) \ exp(-iap_z)$$

$$F_y(p) = \frac{\mp ipp_x - p_yp_z}{p(p_x^2 + p_y^2)} G(p_x, p_y) \ exp(-iap_z) \qquad (5,32)$$

$$F_z(p) = \frac{1}{p} G(p_x, p_y) \ exp(-iap_z)$$

where G is an arbitrary function of p_x and p_y.

Bibliography.

[Bac3]   H. Bacry, J. Phys. *A14*, L73 (1981).

[Bac4]   H. Bacry, *The Position Operator Revisited,*Ann. Inst. H.Poincaré, to appear.

[B.B]   A.O. Barut and A.J. Bracken, Lett. Math. Phys. *7*, 407 (1983).

[Bia]   I. Bialynicki-Birula and Z. Bialynicka-Birula, *Quantum Electrodynamics* (Pergamon Press, Warsaw, 1975).

[B.L2]   L.C. Biedenharn and J.D. Louck, *The Racah-Wigner Algebra in Quantum Theory,* pp.201-221 (Addison-Wesley, Reading, Massachussets, 1981).

[J.J]   A.Z. Jadczyk and B. Jancewicz, Bull. Acad. Pol. Sc., *21*, 477 (1973).

[L.L]   L.D. Landau and E.M. Lifshitz, *Quantum Mechanics* (Pergamon Press, Oxford, 1959).

[L.W.P]   H. Lipkin, W.I. Weisberger and M. Peshkin, Ann. Phys. *53*, 203 (1969).

[Per]   A. Peres, Phys.Rev. *167*, 1449 (1968).

[Pes]   M. Peshkin, Ann. Phys. 66, 542 (1971).

[Schw2]   J. Schwinger, *Particles, Sources and Fields* (Addison-Wesley, Reading, Massachussets, 1970-73).

CHAPTER 6

DEPARTING FROM NEWTON-WIGNER-WIGHTMAN AXIOMS.

Up to now, my point of view was to pretend that there is a position operator for
ch kind of particle. In declaring that the photon does have a position operator, I did not
to affirm that all particles were localizable. With the operator introduced in the
:vious chapter, we know that the photon - as the neutrino - is not localizable, that is,
cannot measure its three coordinates simultaneously. If the reader accepts this idea, he
ll agree with me that we cannot reject the NWW axioms for massless particles alone;
:re are two reasons for that: a) there must be some democracy among particles; this
plies that we must have a unique set of axioms for the construction of the position
erator; b) since one can go continuously from an irreducible representation of the
incaré group $\wp$ corresponding to a massive particle, to the one associated with a
issless particle, it is natural to require continuity for the corresponding position
erators.

Our definition for a position operator will use a given representation of $\wp$. Eq.(5,8)
ative to the photon suggests a solution. It reads

$$K = \frac{1}{2}(H\,R + R\,H\,)\qquad(6,1)$$

ce the energy operator H for the photon is p. It is a simple matter to prove, from
.(6,1) and the commutation relation

$$[R, H] = i\frac{P}{H}\qquad(6,2)$$

t R can be expressed in terms of H and K alone, that is in terms of generators of the
incaré group. We have

$$R = \frac{1}{2}(H^{-1}K + KH^{-1})\qquad(6,3)$$

It is this definition I propose to adopt for the general position operator of any
mentary free system. Let us examine immediately the consequences of such a
inition for a massive particle. For that purpose, it is worthwhile to use the so-called

Foldy canonical expression of the generators of the Poincaré group in terms of the Newton-Wigner operator q, the momentum p and the spin operator s. This is given by

$$P = p \tag{6,4}$$

$$H = \sqrt{p^2 + m^2} \tag{6,5}$$

$$J = q \times p + s \tag{6,6}$$

$$K = \frac{1}{2}(Hq + qH) + \frac{p \times s}{H+m} \tag{6,7}$$

for an irreducible representation. We readily see that, for a spinless particle, the "new" position operator is identical with the Newton-Wigner one q. It follows that every spinless particle is localizable, whatever is its mass value.

The situation is different for spinning particles. We already saw it for massless particles; for massive ones, our definition (6,3) gives

$$R = q + \frac{p \times s}{H(H+m)} \tag{6,8}$$

and it is unlikely that the components are commuting. In fact, we have

$$[R_1, R_2] = - i H^{-3} (m \, s_3 + \frac{p_3 \, p.s}{H+m}) \tag{6,9}$$

and we arrive at an important consequence, namely, *the spinning particles are not localizable*. Since all stable particles are spinning, the non localizability appears as a very fondamental property. Moreover it is a relativistic property as it appears when we write explicitly the fundamental constants c and h in Eq.(6,9). We get

$$[R_1, R_2] = - i \, \hbar c^4 \, H^{-3} (m s_3 + \frac{p_3 \, p.s}{H+mc^2}) \tag{6,10}$$

Eqs (6,8) and (6,10) are easily understood if we take the Galilean limit (c going to infinity), the non relativistic limit (small linear momentum) and the ultrarelativistic limit (large linear momentum).

a) Galilean limit:

$$R \sim q \tag{6,11}$$

$$[R_1, R_2] \sim 0 \tag{6,12}$$

b) <u>Non relativistic limit</u>: $(H \sim mc^2)$

$$R \sim q + \frac{p \times s}{2m^2c^2} \qquad (6,13)$$

$$[R_1, R_2] \sim - i\hbar \frac{s_3}{m^2c^2} \qquad (6,14)$$

and, for the corresponding uncertainty relation,

$$\Delta R_1 \, \Delta R_2 \geq \frac{\hbar}{m^2c^2} |<s_3>| \qquad (6,15)$$

Suppose that we have an electron polarized in the third direction. In that case, we would have

$$\Delta R_1 \, \Delta R_2 \geq \frac{\lambda^2}{2}$$

$$\Delta R_2 \, \Delta R_3 \geq 0 \qquad (6,16)$$

$$\Delta R_3 \, \Delta R_1 \geq 0$$

where λ denotes the Compton wave length.

c) <u>Ultrarelativistic limit</u>:

Now $H \sim pc$. If we denote by n the unit vector in p direction, we get

$$R \sim q + \frac{n \times s}{p} \qquad (6,17)$$

$$[R_1, R_2] \sim - i\hbar n.s \frac{p_3}{p^3} \qquad (6,18)$$

If the particle is in a state of helicity $s = n.s$, Eq.(6,18) coincides exactly, as it must be, with the one of a massless particle (see Eq.(5,9)). The same limit is obtained when the mass m is vanishing in Eq.(6,10).

It is interesting to write the commutation relations in the case where the particle is in a momentum eigenstate (p in the third direction)

$$[R_1, R_2] = -i\,\hbar c^2 \frac{s_3}{H^2}$$

$$[R_2, R_3] = -i\,\hbar c^4 \frac{ms_1}{H^3} \qquad (6,19)$$

$$[R_3, R_1] = -i\,\hbar c^4 \frac{ms_2}{H^3}$$

It is interesting to know at which value of the energy the correction R - q is maximum. According to Eq.(6,8), this corresponds to the maximum of $\frac{p}{H(H+m)}$ or of its square $\frac{H - m}{H^2(H + m)}$. A simple calculation shows that the maximum is reached for the value $H = \phi\, m$, where ϕ is the golden number: $\phi = \frac{1 + \sqrt{5}}{2} \sim 1.618;$ such a situation is a relativistic one[1] (it corresponds to an electron of speed $.786$ c and total energy $.827$ MeV).

The spin-orbit coupling.

One of the best arguments in favour of the operator R is the way we can derive the expression of the spin-orbit energy. For a non relativistic spinning particle, it seems natural to replace the spherical potential $V(q)$ by $V(R)$ in the two-component Schrödinger equation. We have, up to the first order,

$$R^2 \sim q^2 + \frac{L.s}{m^2 c^2}, \quad R \sim q + \frac{L.s}{2m^2 c^2 q} \qquad (6,20)$$

where $\mathbf{L}$ is the orbital angular momentum, and

$$V(R) \sim V(q) + V'(q) \frac{L.s}{2m^2 c^2 q} \qquad (6,21)$$

the corrective term is nothing else than the *spin-orbit energy*.

The Schrödinger position operator for the Dirac equation.

In 1928, when Dirac proposed his relativistic equation for the electron, this equation was considered as a relativistic generalization of the Schrödinger equation, that is as an equation obeyed by the wave function describing the state of a *unique* electron. Many difficulties were present and they only disappeared when the Dirac equation was

---

[1]As underlined by A. Grossmann, such an energy is moderate since it cannot permit a pair emission.

reinterpreted in the framework of quantum field theory. In 1930, Schrödinger made a proposal in order to solve two of these difficulties [Sch1, p.394]. His proposal was forgotten, except one of the byproducts known as the *zitterbewegung*.

The two difficulties were the following ones:
i) the presence of negative energy states[2]
ii) the fact that any component of the speed had only two eigenvalues, namely $\pm c$. Indeed,

$$H = c\alpha.p + \beta mc^2 \qquad (6,22)$$

and

$$\frac{dX}{dt} = -i[X, H] = c\alpha_x \qquad (6,23)$$

Schrödinger proposed to replace his own non relativistic position operator X by

$$X_S = \Pi_+ X \Pi_+ + \Pi_- X \Pi_- \qquad (6,24)$$

where the operators $\Pi_\pm$ are the Hermitian projections onto the space of positive (resp. negative) energy states.

It is obvious that if the potential is expressed in terms of X_S rather than in terms of X, the time evolution operator of the Dirac equation will keep a positive energy state in the space of positive energy states. That solves the first difficulty. Moreover, we have

$$\frac{dX_S}{dt} = -i[X_S, H] = \frac{c^2 p_x}{H} \qquad (6,25)$$

which is a satisfactory result.

The Schrödinger proposal is discussed here because the operator X_S coincides exactly with the operator R defined by Eq.(6,3) for the representation of $\wp$ corresponding to the Dirac equation. The space of solutions of the Dirac equation is the direct sum of two irreducible subspaces under $\wp$. The projections associated with these two subspaces are $\Pi_\pm$. The operator R of Eq.(6,8) is nothing else than the operator

$$\Pi_+ X \Pi_+$$

---

[2]Let us quote Schrödinger himself; we are in 1930, two years before the discovery of the positron:*"Les valeurs propres négatives [de l'énergie] n'ont pas de signification physique; on voudrait bien s'en débarrasser. Au moins il devrait être impossible qu'une fonction propre "positive" se transforme au cours du temps en donnant naissance à des fonctions "négatives" ou tout au moins cette variation ne devrait se produire qu'infiniment lentement pour rendre suffisamment improbable l'énorme changement d'énergie $2mc^2$ que nous n'avons jamais observé"* [Sch1,vol.3]. We must also underline that the presence of negative energy states is not a quantum fact. As Dirac emphasized it, they are not discarded explicitly by the classical theory of special relativity.

that is the restriction of the operator X to the space of positive energy states.

The *zitterbewegung* is described by the vector operator $\xi = X - X_S$. We note that there exist *a priori* many ways of writing explicitly the vectors ξ and X_S; it could be worthwhile to explain why. We are interested in operators acting on the space D of solutions of the Dirac equation; this is a subspace of the space S of all spinor functions on space-time on which the operators α, X, p are defined. The restrictions of these operators to the space D are usually denoted by the same letters for the sake of simplicity. However, it is clear that there are many operators acting on S such that their restriction on D is X_S. In the case of the massless particles, the arbitrariness of the "vector potential $A(-p)$" has the same interpretation.

To conclude, let me give the expression of the position operator for a particle with spin $\frac{1}{2}$ governed by the two component Weyl equation

$$(p_0 - \sigma.p)\psi = 0 \tag{6,26}$$

It is given by [Bac4]

$$R = \Pi \, X \, \Pi \tag{6,27}$$

with

$$\Pi = \frac{1}{2}\left(1 + \frac{\sigma.p}{p}\right) \tag{6,28}$$

Bibliography.

[Bac4]   H. Bacry, *The Position Operator Revisited,* Ann. Inst. H. Poinc. (to appear).
[Fol]   L. Foldy, Phys.Rev. *102,* 568 (1956).
[Sch1]   E. Schrödinger, *Collected Papers* (Austrian Academy of Science, Vienna, 1984).

CHAPTER 7

MINKOWSKI SPACE-TIME SUITABLE FOR PARTICLE PHYSICS?

> *"There is the definite possibility that some future theory may be found which describes nature more accurately than present theory, but for which the differentiable manifold picture of space-time would not be appropriate...I do not believe that a real understanding of the nature of elementary particles can never be achieved without a simultaneous deeper understanding of the nature of space-time itself."* (R. Penrose, in [D.W]).

> *"In particle dynamics the dynamical object is not* x *and* t, *but only* x... *This understood, how can physicists change their minds and "take back" one dimension? The answer is simple. A decade and more of work by Dirac, Bergmann, Schild, Pirani, Anderson, Higgs, Arnowitt, Deser, Misner, de Witt and others has taught us through many a hard knock that Einstein's geometrodynamics deals with the dynamics of geometry: of 3-geometry, not 4-geometry."* (J.A.Wheeler, in [D.W]).

Let us sum up the results of the previous chapters. Each elementary system, in the
igner sense, has been attributed a position operator. For spinning elementary particles
 particular, the stable ones: proton, electron, neutrino, photon), the components of this
erator are not commuting. In support of such an operator, we have obtained the three
lowing arguments:

a) every particle has a position operator; therefore, it makes sense to speak of a coordinate measurement even for massless particles;

b) the electron slit experiment has a pure quantum interpretation in the standard statistical description of quantum mechanics;

c) in the non relativistic limit, that is for low momenta, the spin-orbit coupling comes out from the replacement of the Schrödinger operator X by the new one R.

Up to now, we have only proposed to give up the old Schrödinger position operator and to replace it by a more sophisticated one. We could be satisfied by this point of view and declare that it is the end of the story. Unfortunately (or fortunately) this is not possible because many conceptual difficulties are still present and it is somewhat easy to see that adopting our new operator has very many important consequences[1]. In particular, since it is impossible to measure a sharp position for a stable particle, it becomes difficult to build Minkowski space-time from the set of quantum states of the photon; in other words, there is no way to rediscover the X operator from the Bias D operator. However, we are authorized to think that in a world where we only have photons, we must be able to obtain a description of Minkowski space-time for the same reason Einstein was able to build it from light signals. In this context we have already underline that Einstein ignored in his construction the transverse character of light waves, a property which is at the origin of the non localizability of the photon (both in the Newton-Wigner-Wightman sense and in the sense we have adopted). The necessity of being able to build space-time from quantum physics is completely in the spirit of quantum ideas; as says Wigner: *"Quantum mechanics told us that we should describe situations or events only in terms of quantities that can be observed"*[Wig3]. This implies that we have to use quantum observables and, especially in this case, the operator D.

Then we are led to bring into question the reality of Minkowski space-time itself, I mean as an ingredient of quantum physics. The paradox is that Minkowski space-time is intimately related with covariance, that is with the Poincaré group and it is the Poincaré group which led us gradually to consider seriously the question of giving up the Minkowski space-time. Moreover, the conservation of energy-momentum, which is a byproduct of the Poincaré group invariance, is so well satisfied that none particle physicist would be ready to give up the Poincaré group[2].

---

[1] as announced by Wightman:*"I venture to say that any notion of localizability in three-dimensional space which does not satisfy [my axioms] will represent a radical departure from present physical ideas"*[Wig].

[2] Schrödinger considered the possibility of giving up the Poincaré group: *"Il apparaît donc qu'au point de vue de la mécanique quantique, la théorie de la relativité se range au même niveau que la mécanique classique en ce sens qu'elle ne représente qu'une approximation relative au domaine macrocospique. On ne devra pas admettre tout simplement les formules de la relativité (par exemple les formules de Lorentz) et les supposer valables sans changement dans le domaine intra-atomique. Elles devront être soumises à des modifications qui seront probablement analogues à celles qu'a subies la mécanique ordinaire pour se transformer en mécanique quantique. Il faudra "quantifier" la transformation de Lorentz"*[Sch, vol.3, p.417]. One of the arguments of Schrödinger is the fact that Einstein clocks must be infinitely heavy if we want them to provide us with a sharp measurement of time.

Fortunately, there is no contradiction in keeping the Poincaré group and the linear energy-momentum space and rejecting simultaneously the Minkowski space-time. These two four dimensional spaces are quite different, both from physical and mathematical point of view. We give, in Table 1, the main differences between these two spaces.

<u>Minkowski space</u>	<u>Energy-momentum space</u>
- Homogeneous space (all points are "equal"). The Poincaré group acts transitively on this space.	- Not homogeneous. The Poincaré group acts on orbits (mass shells).
- Affine space.	- Vector space.
- can be identified with the quotient Poincaré/ Lorentz.	- Dual of the translation subgroup.[3]
- Dimension: length.	- Dimension: action/ length.
- Elements appear as dummy variables[4] in QFT calculations where they loose their space-time meaning[5].	- Conserved quantities (easily measured since transferable to apparatuses). They are also additive quantities[6].
- The only use in QFT lies in writing the locality of the interaction Lagrangian.	- All calculations of Feynman diagrams are made in this space[7].

<u>Table 1</u>

One is tempted to say that the role of Minkowski space was essentially a historical one, in contradistinction to the energy-momentum space which is fundamental in particle physics. The two spaces are so intimately related (or sometimes identified!) in physicist's

---

[3]It could also be considered as a subspace of the dual of the Lie algebra of $\wp$.

[4]They are just integration variables in the action integral.

[5]as underlined by Sakurai:*"It is important to note that the x and t that appear in the quantized field A(x,t) are not quantum-mechanical variables but just parameters on which the field operator depends. In particular, x and t should not be regarded as the space-time coordinates of the photon"*[Sak,p.32].

[6]Very often, people fail to tell the difference between "conserved" and "additive". Since the energy-momentum of an *isolated* system is conserved, its mass, which is a function of the energy-momentum, is also conserved. However, the mass is not additive.

[7]Mandelstam variables are functions of the energy-momentum, a picture in a bubble chamber is analyzed in this space, etc.

minds that a real effort is needed to convince oneself that it is possible to give up only one of them. That Minkowski space-time is not satisfactory for the particle physicist is very often underlined as we are going to show[8]. First, as underlined by *Schwinger, Minkowski space-time is probably responsible of the divergences of QED: "We conclude that a convergent theory cannot be formulated consistently within the framework of present space-time concepts"* [Schw1, preface]. But the main conceptual difficulty lies in the contradiction between the localizable field and the non localizability of particles. Let us examine this point.

"A localizable dynamical variable is a quantum-mechanical operator which describes the physical conditions at one particular point x in space and time". Every quantum field theoretician would agree with that definition taken from Jauch and Rohrlich's textbook[J.R]. The problem is that if we want to analyze further this notion, we are readily led to contradictions. Indeed, the same authors write: *"A localizable dynamical system is one for which a complete set of localizable dynamical variables exists. Thus, for instance, a system of point particles is a localizable system, the localizable variables in this case being the position of the particles and their spin, if any"*. We readily see that the concept of a localizable dynamical variable is in complete contradiction with the Newton-Wigner result about the photon and the neutrino.[9] Schwinger is aware of the difficulty since he writes:

"A localizable field is a dynamical system characterized by one or more operator functions of the space-time coordinates, $\phi^\alpha(x)$. Contained in this statement are the assumptions that the operators x_m, representing position measurements, are commutative,

$$[x_m, x_n] = 0,$$

and furthermore, that they commute with the field operators,

$$[x_m, \phi^\alpha] = 0,$$

so that

$$(x \mid \phi^\alpha \mid x') = \delta(x-x') \, \phi^\alpha(x).$$

The difficulties associated with current field theories may be attributable to the implicit hypothesis of localizability"[Schw1, p.343].

Do not we have to conclude with Schrödinger: *"Mais l'espace géométrique est une création de notre imagination, qui s'est développé par et pour l'usage de la vie de tous les*

---

[8]I was surprised to read in [Dre, p.211]: *"Bohr's main conclusion from the failure of the BKS theory was that a profound, radical revision of all the physical concepts had now become inevitable. Only by a complete renunciation of the usual space-time methods of visualization of the physical phenomena would further progress become possible. The impossibility of maintaining the usual causal space-time description was a recurring theme in Bohr's thinking; it eventually culminated in the* complementary *formulation of quantum mechanics"*. Do we have to conclude, from this source, that Bohr *also* was not satisfied by Minkowski space-time?

[9]From our point of view, the concept of a point particle is even meaningless, at least if it is spinning.

jours. Il n'est nullement sûr que la nature possède les propriétés de cette construction mentale"[Sch1, vol.4, p.370]?[10] or to say, as Wigner, *"How can space-time points be defined? This is a difficult question...I believe our present idealized space-time concept will undergo modification"*[Wig3].

"What is the significance of the variables x_μ...? A priori there is no reason for the x_μ's to have anything to do with space and time." [K.W]. This is probably a little too drastic as a statement, but it is significantly a proof that many physicists are troubled by the meaning of the x_μ's. Many people would agree that they are just *labels* (what are they labelling?). Less are aware that they are just dummy variables. We have to accept the inevitable fact that *they do not correspond to quantum observables* . Their link to space and time is rather mysterious and it is not natural to accept such a situation without trying to understand it; my deep conviction is that there must be a way to derive the classical Minkowski space-time from quantum field theory *and not the converse* as we are taught by quantum field theory. One could be tempted to object that time, in contradistinction with position, is not an observable but just a parameter in ordinary quantum mechanics and that nobody complains about it. That is not true. As Schrödinger emphasizes it, *time is measured* ; a time measurement answers the question: *"what time is it* now?*"*[11]

The impossibility of measuring simultaneously the three coordinates of a particle has another important consequence: the quantization procedure looses completely its value. The quantum observables R_i we have introduced have commutation relations which have nothing to do with the Poisson brackets of classical mechanics. The path to follow from classical to quantum mechanics has necessarily an intermediate step corresponding to some modification of our concept of space. Another important consequence is that if we accept the impossibility of quantizing worldlines (one-dimensional submanifolds in Minkowski space-time), we could not see, *a fortiori*, why we would be authorized to quantize strings which are two-dimensional submanifolds. This remark makes highly improbable the possibility of understanding strong interactions with the aid of quantum strings. However, it is interesting to mention that the *R* operator introduces an effect which reminds strings. Indeed, there are situations where we are able to measure with a quite good precision two of the three coordinates of a particle; in such a xase, due to the uncertainty relations, the third coordinate has a very fuzzy value and the particle looks like a string *but keeps its usual number of degrees of freedom*. Of course, in order to look like a string, we must have sharp values of two coordinates, a condition which depends on the spin state of the particle.

Another remark is of interest: since, according to our proposal, a spinning particle does not have a classical *point* image, it can be expected that this property will permit to escape the infinity of the self-energy of the particle. This could be counted as a fourth argument in favour of our position operator. After all, the notion of point in a continuum

---

[10]It is paradoxical that Schrödinger who was strongly in favour of the fundamental role of the wave function - a function on the ordinary space! - was ready to give up this object.

[11]The reader will convince himself that such a question corresponds to a situation where we have really *two* clocks, one for the measurement, the other for defining "now". I hope the reader will forgive me not to enter the problem of time in the quantum theory of measurement: everybody knows that any measurement takes time...

iis intimately related with the notion of infinity (how to define real numbers without the concept of infinity?); it is not surprising that divergences are present when *point particles* are welcome[12].

We have mentioned in Table 1 that the only role of space-time variables is to write the locality of interactions. That it is true can be seen in the following way. Suppose that we are interested in a system of photons, electrons and positrons. As far as these particles are not interacting, we have no need of field theory: we start with the Hilbert spaces of the one photon (resp. electron or positron) states; they are just given by the representation spaces of the Poincaré group, i.e. without the help of Minkowski space-time; then, with the only aid of the tensor notion, we build the Fock spaces associated with the corresponding "fields" *without using the notion of field* [13](space-time is not involved in such a construction); then we introduce the annihilation and creation operators with the usual commutation or anticommutation relations; if the "fields" are not interacting, the Hamiltonian is just given by the generator of the time translations of the Poincaré group, an operator which is known since we started with the representations associated with each kind of particles. If the "fields" are interacting, we have to choose a Hamiltonian in terms of the creation and annihilation operators. The locality of interactions is obviously a way of making a selection between all possible lagrangians but the interaction term is directly taken from classical physics although we are in the so-called *second quantization* procedure. It is strange that we do not take into account the position operators appearing in the *first quantization* part of the theory. We must also underline that in first quantization, position and time do not play analogous roles since one is an operator, the other one is just a parameter. It is surprising that in the next step (second quantization), we go back to the classical situation with the four x_μ as parameters! Do we really believe in what we are teaching, namely *the existence of a position operator*? Moreover, as it has been emphasized by Connes, what is the value of the locality of interactions requirement when it is really forgotten after the renormalization of the theory?

It is possible to quote other people about conceptual difficulties encountered in quantum field theory. For instance, *"Field theory works fine for free fields but the ideas don't seem to carry over very well for the interacting case"*[Sta] or, from the same author, *"A(x) that first appears in Lagrangian field theory... is defined for points x rather for wave packets of freely moving particles. This original A(x) is found to be a rather unsatisfactory object because its matrix elements between physical states are all zero. Thus it is multiplied by infinity and one gets the renormalized field..."*

The reader would have understood that I am trying to convince him to give up the Minkowski space-time and, consequently, the notion of quantum field defined on it. The problem is of course to find a substitute for the field concept; my opinion is that space-time must be quantized in some way. If I am right, that could imply that the gravitational

---

[12]It is really surprising that many authors are referring, in books on quantum mechanics, to *point particles*. They are usually very prudent in not trying to make this expression more precise. Speaking of point particles supposes implicitly that our apprehension of microscopic structure of space is quite clear. Obviously, it is not.

[13]It would be nice to propose a new name for a "field" which does not refer to a continuum.I keep the word field between quotation marks. A possible name would be a *Fock field*.

field has no quantum equivalent and that would explain why it is so difficult to quantize it. After all, up to now, this field is known as a *macroscopic* field and quantum mechanics describes essentially *microscopic* phenomena.

We have to underline that our position operator was defined for a *free* particle. However we have used it in the Schrödinger equation and obtained the spin-orbit coupling. I must explain why I do not have to do the same thing in the case of the Dirac equation with an external potential. It is not enough to say that the spin-orbit coupling is directly obtained with the use of the standard potential $V(X)$. Introducing a potential in the Dirac equation is highly not natural. It would imply that we are considering this equation at the first quantization level. Clearly, the Schrödinger equation is the quantum equivalent of the classical equation of motion of a Hamiltonian system; we do know the role played in this framework by a potential; the situation is quite different in classical special relativity where there is no room for potentials. This is well known but, nevertheless, we know - and it is a miracle - that the Dirac equation provides a rather good approximation for the hydrogen spectrum[14]. Let me emphasize once more what were the concepts introduced by special relativity: the Poincaré invariance and its corollary, the conservation of energy-momentum; the Minkowski space-time is unable to describe interactions between particles.

Before concluding, I would like to invite the reader to meditate a recent article by Wigner [Wig3] from which I extract the following text which has the advantage to enlarge the problem in including general relativistic aspects:*"How can space-time points be defined? This is a difficult question and, as we will see, it also plays an important role outside the general theory of relativity. But in general relativity, it is a basic question.*
"In classical theories, space-time points are best defined as crossing points of the paths of two objects - naturally infinetely small ones. And in general relativity, it is implicitly assumed that there are infinitely many such very light objects, so that the intersections of their worldlines define a sufficiently dense set of space-time points. This is, evidently, a very wild assumption and one must admit that the general relativity theory is not really positivistic.
"The situation is worse in quantum mechanics. The objects have no paths and the coincidence of two is not defined - there is no "point of collision." The collision matrix, which can be determined by many repeated experiments, does not define the point of collision. It is implicitly assumed, both in general relativity and in quantum mechanics, that there are macroscopic measuring systems which enable the determination of the coordinates of spce-time points, but the influence of these systems on the systems under observation can be neglected. Altogether, as will be discussed further,the real existence[15] of space-time points and the possibility of determining their coordinates is an assumption both in general relativity theory and in quantum mechanics - particularly in the field theories of the latter - but is very questionable in both. I believe that even the probability of the system's particles to have given positions at definite times is not determinable - the magnitude of the field strengths at a space-like surface even less. The point will be further supported below. Its realization will, I believe, fundamentally change our quantum

---

[14]under a strange condition: we have to replace the electron mass by the reduced mass, a concept which has no room in special relativity!
[15]That is probably the only reference to a metaphysical question in the present book.

mechanics and probably all fundamental concepts of our physics."

I arrived at the end of this work. It is probably clear to the reader that if I was able to write the "next chapter" of my book, I would have written it. This does not mean that I do not know in which direction I have to look in order to go further. Let me say a few words about that. If we give up the Minkowski space-time as a continuum, it does not follow necessarily that we have to replace it by a discrete object. That would lead us to the desertion of the continuous Poincaré group which is, as I underlined it, a very useful ingredient for particle physics. The mathematicians taught us that all information concerning a manifold can be obtained from the study of the functions on it[16]. These functions form a commutative algebra. Conversely, given any (not necessarily commutative) algebra, one can associate with it a (not necessarily "commutative") manifold. There is a possibility that our space-time is such an object...

A last comment is necessary. In all chapters, I was essentially concerned with the notion of space *alone*. What about time? From Newton-Wigner's results, we know that it is not possible to find a position operator having the Lorentz covariance property (remember, for instance, that a state cannot be a localized one for two distinct observers). But, even if such an object was possible, it would be difficult to conciliate the operator character of the position with the parameter character of time. It seems more natural to accept this difference as an *a priori* fact: first, let us understand space...

Bibliography

[Co]   A. Connes, Pub. Math. I.H.E.S., *62*, 257 (1986).
[D.W]   C.M. DeWitt and J.A. Wheeler, *1967 Lectures in Mathematics and Physics* (Battelle Rencontres, Benjamin, New York, 1968).
[Dre]   M. Dresden, *H.A. Kramers, Between Tradition and Revolution* (Springer-Verlag, New York, 1987).
[J.R]   J.M. Jauch and F. Rohrlich, *The Theory of Photons and Electrons* (Addison-Wesley, Reading, Massachussets, 1955).
[K.W]   V. Kaplunovsky and M. Weinstein, Phys.Rev. *D31*, 1879 (1985).
[Sak]   J. Sakurai, *Advanced Quantum Mechanics* (Addison-Wesley, Reading, Massachussets, 1967).
[Sch1]   E. Schrödinger, *Collected Papers,* vol. (Austrian Academy of Science, Vienna, 1984).
[Schw1]   J. Schwinger, *Quantum Electrodynamics* (Dover, New York, 1958).
[Sta]   H.P. Stapp, *Lectures on the S-matrix theory,* University of California, Lawrence Radiation Lab., Berkeley (1961).
[Wig]   A.S. Wightman, Rev. Mod. Phys. *34*, 845 (1962).
[Wig3]   E.P. Wigner, Int. Jour. Theor. Phys. *25*, 467 (1986).

---

[16] The physical counterpart of it lies, for instance, in the fact that the physicists are able to distinguish between a point electric charge from a point magnetic monopole in space and that they know how to relate them to the singularities in space to Maxwell's equations.

APPENDIX A

SYMPLECTIC STRUCTURE OF COADJOINT ORBITS.

Let G be a Lie group and G its Lie algebra. We denote by G^* the dual vector space of G. If $x, y \in G$ and $f \in G^*$, we have

$$< coad(y)f, x > \; = \; < f, ad(y) x > \; = \; < f, [x, y] >$$

Denote by G_f the Lie subalgebra stabilizing f; then G/ G_f is the tangent space to the coadjoint orbit and the symplectic form is, for $X, Y \in G/ G_f$,

$$(X, Y)_f = \; < f, [X, Y] >$$

which is clearly antisymmetric.

APPENDIX B

QUANTIZATION OF THE SPHERE S^2.

The principle of the quantization procedure is the following one. Given a symplectic manifold M with σ as a symplectic form[1], one considers a covering of M by simply connected open sets M_i such that the restrictions σ_i are integrable: $\sigma_i = d\varpi_i$ (d denotes the exterior derivative) and satisfy the compatibility conditions:

$$\varpi_i\,(dx) - \varpi_j\,(dx) = dz_{ij}\,/\,i\,z_{ij}$$

The symplectic form of S^2 can be written as a mixed vector product:

$$\sigma_{\mu\nu}(r)\,dr^\mu\,\delta r^\nu = \lambda\,(r,\,dr,\,\delta r)$$

where λ is a given real positive number. By performing a stereographic projection from North ($\varepsilon = 1$) or South ($\varepsilon = -1$) pole, we get

$$x = (1+\alpha^2+\beta^2)^{-1}\,2\alpha r,\quad y = (1+\alpha^2+\beta^2)^{-1}\,2\beta r,\quad z = \varepsilon(1+\alpha^2+\beta^2)^{-1}(1-\alpha^2-\beta^2)\,r$$

A simple calculation shows that

$$\sigma = 4\varepsilon\lambda(1+\alpha^2+\beta^2)^{-2}\,d\alpha\wedge d\beta = 2\varepsilon\lambda\,d[(1+\alpha^2+\beta^2)^{-1}(\alpha d\beta-\beta d\alpha)] = d\varpi_\varepsilon$$

The stereographic projections map two open sets (the Northern and Southern hemispheres) on $\mathbb{R}^2$. The quantization condition reads:

$$\sigma_+ - \sigma_- = dz/\,iz$$

which gives by integration

---

[1] If x denotes a point of the manifold M, we have an antisymmetric tensor field $\sigma_{\mu\nu}(x) = -\,\sigma_{\nu\mu}(x)$ satisfying $det\,\sigma_{\mu\nu} \neq 0$.

$$z = z_0 \, exp(2i\lambda\phi), \quad (\phi \text{ is the longitude})$$

Therefore 2λ is an integer.

Bibliography

[Sou2]   J.-M. Souriau, *Structure des systèmes dynamiques* (Dunod, Paris, 1970).

APPENDIX C

THE ANGULAR MOMENTUM OF AN ELECTRIC CHARGE IN THE MONOPOLE FIELD (AN APPLICATION OF THE NOETHER THEOREM).

Let us consider a classical point particle of mass m and electric charge e in the field B of a monopole with magnetic charge g. We have

$$B(r) = g \frac{r}{r^3} \qquad (C,1)$$

We denote by $A(r)$ the vector potential (necessarily singular) defined by

$$B(r) = g \; curl \; A(r) \qquad (C,2)$$

(we put the factor g for convenience).

The Hamiltonian of the problem is

$$H(r,p) = \frac{(p - egA(r))^2}{2m} \qquad (C,3)$$

and the Lagrangian reads

$$L(r,v) = \frac{1}{2} mv^2 - egA(r).v \qquad (C,4)$$

Since the field (C,1) is invariant under rotations, Noether's theorem will provide us with a conserved angular momentum . To get its expression, we perform an infinitesimal rotation around the third axis:

$$\delta x = - y\phi, \qquad \delta v_x = - v_y\phi$$

$$\delta y = x\phi, \qquad \delta v_y = v_x\phi \qquad (C,5)$$

$$\delta z = 0, \qquad \delta v_z = 0$$

We get

$$\delta L(r,v) = - eg(\frac{\partial A_x}{\partial x} \delta x + \frac{\partial A_x}{\partial y} \delta y)v_x$$
$$- eg(\frac{\partial A_y}{\partial x} \delta x + \frac{\partial A_y}{\partial y} \delta y)v_y \qquad (C,6)$$
$$- eg(\frac{\partial A_z}{\partial x} \delta x + \frac{\partial A_z}{\partial y} \delta y)v_z$$
$$- eg(A_x \delta v_x + A_y \delta v_y)$$

In using Eqs(C,2) and (C,5) alone (that is without any explicit expression of the vector potential), we obtain easily

$$\delta L(r,v) = eg\phi\frac{d}{dt} (A_x y - A_y x - \frac{z}{r}) \qquad (C,7)$$

and the Noether conserved quantity is given by

$$\phi J_z = \frac{\partial L}{\partial v_x} \delta x + \frac{\partial L}{\partial v_y} \delta y + eg\phi (A_x y - A_y x - \frac{z}{r}) \qquad (C,8)$$

or

$$J = r \times (p - egA) - eg\frac{r}{r} \qquad (C,9)$$

APPENDIX D

NON LOCALIZABILITY OF THE CLASSICAL MASSLESS PARTICLE WITH HELICITY.

In the present appendix, we want to show that the non localizability of the massless particle with helicity is only due to special relativity, without the use of quantum theory.

The problem is to define a wordline for a classical particle which has energy-momentum P_μ and generalized angular momentum $M_{\mu\nu}$ such that the Pauli-Lubanski vector W_μ obeys

$$W_\mu = s\, P_\mu, \tag{D,1}$$

where s is the (non zero) helicity. Due to the orthogonality of the 4-vectors P and W, Eq. (D,1) implies that these vectors are light-like and, therefore, that the particle has zero mass. This equation has a non covariant writing. Denoting by J and $-K$ the "magnetic" and "electric" parts, respectively, of the tensor $M_{\mu\nu}$, we get

$$W_0 = J \cdot P = s\, P_0 \tag{D,2}$$

$$W = P_0\, J - P \times K = s\, P \tag{D,3}$$

Let us decompose the angular momentum J in transverse and longitudinal parts with respect to P:

$$J = J_\parallel + J_\perp \tag{D,4}$$

Eqs (D,2) and (D,3) give

$$s = J_\parallel \tag{D,5}$$

$$P_0\, J_\perp = P \times K \tag{D,6}$$

In order to associate a worldline with the generalized momentum $(M_{\mu\nu}, P_\mu)$, we have to perform a translation which would provide $M_{\mu\nu}$ with some "canonical property", as it is the case in classical mechanics: the angular momentum of a point particle is minimum (zero) with respect to the point where the particle lies. Under a space translation

T, the angular momentum J transforms in the following way

$$J \rightarrow J - T \times P \tag{D,7}$$

It follows that we can choose T in order to cancel the transverse part of J. A translation has no effect on the longitudinal part (in fact, we know that s is an invariant). In that case, Eq. (D,6) shows us that K is collinear to P. These results can be written

$$J = s \frac{P}{P_0} \tag{D,8}$$

$$K = a P \tag{D,9}$$

Now, it is clear that a time translation permits to cancel K (but does not affect J). What we have shown is that there is a point in Minkowski space-time for which K is zero and J collinear to P. This point is not unique since any light-like translation in the P_μ direction provides us with another point with the same property. In fact, we get in this way a whole (light-like) worldline of such points. It is natural to consider this line as the canonical worldline of the particle. Unfortunately, we are going to show that there is an infinite number of such worldlines!

Let us perform a boost of speed v (we choose $c = 1$). Denoting by γ, as usually, the quantity $(1 - v^2)^{-1/2}$, the transformation formulas for the relativistic momenta associated with the Poincaré group read

$$P'_o = \gamma P_o - \gamma P . v$$

$$P' = P - \gamma P_o v + \frac{\gamma^2}{1 + \gamma} (v.P) v$$

$$\tag{D,10}$$

$$J' = J + \gamma v \times K - \frac{\gamma^2}{1 + \gamma} v \times (v \times J)$$

$$K' = K - \gamma v \times J - \frac{\gamma^2}{1 + \gamma} v \times (v \times K)$$

(the two last formulas are the well known transformation formulas for the electric vector $-K$ and the magnetic vector J).

Let us consider the case where (D,8) is satisfied and K vanishes. We have, for instance,

$$P_o = p$$

$$P = (0, 0, p), J = (0, 0, s), K = (0, 0, 0)$$
(D,11)

According to our hypothesis, this means that the particle is located at the origin. First we note that, if we boost along the third direction, we can make p as small or as large as we want, without modifying J and K.

Let us now perform a boost of speed v in the first direction.

$$v = (v, 0, 0)$$
(D,12)

We get

$$P_o' = \gamma p$$

$$P' = (-\gamma p v, 0, p), J' = (0, 0, \gamma s), K' = (0, \gamma v s, 0)$$
(D,13)

The Lorentz transformation does not preserve our requirement about Eq.(D,8) and the vanishing of K. This is a very serious disease since the Lorentz transformation is not sufficient to provide us with the worldline of the massless particle[1]. In order to reach the required conditions, we have to perform an extra translation

$$T' = (0, p^{-1}vs, 0)$$
(D,14)

Such a translation does not transform P_o' nor P' but gives

$$J' \rightarrow J' - T' \times P' = (0, 0, s)$$

$$K' \rightarrow K - P_o' T' = (0, 0, 0)$$
(D,15)

Eq.(D,15) proves that the trajectory is shifted in the second direction. The shift $p^{-1}vs$ is as large as we want since, as we already noted, p is arbitrary small . It follows that if, for one observer, the particle is located on the earth and moving towards the polar star, there always exists an observer who will locate it near a star located, say, in a zodiac constellation. We must emphasize that this paradoxical result is obtained in the framework of classical special relativity alone and is due, for the photon, to its spin, that

---

[1]It is however sufficient for a spinless particle (with or without mass).

is to the transverse character of Maxwell waves. The non localizability of the photon is not a specific result of quantum relativity.

Bibliography

The reader is referred to the one given in Chapter 3.

APPENDIX E

THE MAXWELL EQUATIONS AND THE POINCARE GROUP.

In the present appendix, we want to relate the Maxwell equations with the Wigner work on projective representations of the Poincaré group. We intend to present two versions of it, one concerning the field itself, the other concerning the potentials. In both cases we will see that what is involved is a representation of the *whole* Poincaré group (i.e. with parity and time reversal).

1. The electromagnetic field.

If we define F as the complex field $B - i E$, we know that the Lorentz group acts as the complex rotation group. More precisely, any Lorentz transformation is uniquely factorized in a product of a rotation by a boost. A rotation of angle ϑ around a unit vector u is given by

$$F' = F + \sin \vartheta \, u \times F + (1 - \cos \vartheta) u \times (u \times F) \tag{E,1}$$

and a boost of speed v is given by the formula[1]

$$F' = F - i\gamma v \times F - \frac{\gamma^2}{1+\gamma} v \times (v \times F) \tag{E,2}$$

where

$$\gamma^2 = (1 - v^2)^{-1} \tag{E,3}$$

It is a simple matter to check that the similar formulas (E,1) and (E,2) are complex rotations since both preserved the "square length" F^2, that is the real part $B^2 - E^2$ and the imaginary part $2B.E$. This shows that the complex rotation group $SO(3,C)$ is isomorphic to the Lorentz group. This action is the "spin" action.

The generators of $SO(3,C)$ are the matrices

---

[1]Eq.(E,2) is the same as the one obeyed by $J + iK$ of Eq.(D,10).

$$S_1 = \begin{pmatrix} 0 & 0 & 0 \\ 0 & 0 & -i \\ 0 & i & 0 \end{pmatrix}$$

$$S_2 = \begin{pmatrix} 0 & 0 & i \\ 0 & 0 & 0 \\ -i & 0 & 0 \end{pmatrix} \tag{E,4}$$

$$S_3 = \begin{pmatrix} 0 & -i & 0 \\ i & 0 & 0 \\ 0 & 0 & 0 \end{pmatrix}$$

for the real rotations and iS_1, iS_2, iS_3 for the boosts[2].

We already mentioned that electrodynamics is a theory which has the whole Poincaré group as an invariance group. Under parity, the vectors B and E transform as follows[3]

$$\text{Parity: } B \rightarrow B, E \rightarrow -E. \tag{E,5}$$

In order to implement parity, we have to double the number of components of the field since F is transformed into F^*. The "spin" generators of the Lorentz group become six dimensional matrices, namely,

$$\Sigma_i = \begin{pmatrix} S_i & 0 \\ 0 & S_i \end{pmatrix}, i\Gamma_5 \Sigma_i = \begin{pmatrix} iS_i & 0 \\ 0 & -iS_i \end{pmatrix} \tag{E,6}$$

where the matrix Γ_5 is defined by

$$\Gamma_5 = \begin{pmatrix} 1 & 0 \\ 0 & -1 \end{pmatrix} \tag{E,7}$$

---

[2]The group $SO(3,C)$ is a three dimensional complex group generated by the three matrices S_i; however, as a real Lie group, it is six dimensional.
[3]Under time reversal, we have $B \rightarrow -B, E \rightarrow E$, the doubling is also necessary.

We note that the parity matrix Γ_0 reads

$$\Gamma_0 = \begin{pmatrix} 0 & 1 \\ 1 & 0 \end{pmatrix} \tag{E,8}$$

If we add the orbital and the "spin" generators of the Poincaré group, we get

$$\boldsymbol{J} = -i\,\boldsymbol{x} \times \nabla + \Sigma$$

$$\boldsymbol{K} = -it\,\nabla -i\boldsymbol{x}\,\frac{\partial}{\partial t} + i\Gamma_5\Sigma$$

$$\boldsymbol{P} = -i\,\nabla \tag{E,9}$$

$$H = i\frac{\partial}{\partial t}$$

These generators describe the action of $\wp$ on *any* three dimensional vector field. We are going to show that if we want such a field to correspond to a massless representation of $\wp$, it will obey necessarily Maxwell equations and its helicity operator will have ± 1 as eigenvalues.

The Pauli-Lubanski vector has components

$$W^0 = \boldsymbol{J} \cdot \boldsymbol{P} = \Sigma \cdot \nabla$$

$$\boldsymbol{W} = \boldsymbol{J}H + \boldsymbol{K} \times \boldsymbol{P} = i\,\Sigma\frac{\partial}{\partial t} + \Gamma_5 \times \nabla \tag{E,10}$$

We note that W^0 is, up to an i factor, the *curl* operator[4].

We want to select a massless representation of helicities $\pm s$. According to Wigner's results, we have to impose the conditions

$$W_\mu - s\,\Gamma_5\,P_\mu = 0 \tag{E,11}$$

which means

---

[4]This writing shows that the curl operator is a scalar operator in Racah's terminology, since it commutes with $\boldsymbol{J}$.

$$(\Sigma . \nabla + s \; \Gamma_5 \frac{\partial}{\partial t}) \; \Phi = 0 \tag{E,12}$$

$$(\Sigma \frac{\partial}{\partial t} - i \; \Gamma_5 \Sigma \times \nabla + s \Gamma_5 \nabla) \; \Phi = 0 \tag{E,13}$$

where Φ is a six components field collecting F and F^*.

Eq.(E,12) is just $curl \; F - is\frac{\partial F}{\partial t} = 0$ and its conjugate. Eqs(E,13) are not all independent. They are compatible provided $s^2 = 1$ (which implies that Eq.(E,12) is a Maxwell equation) and they contained the other Maxwell equation $div \; F = 0$.

The introduction of the Γ–matrices makes explicit the analogy between Maxwell equations and the Dirac equation[5]. In particular, one can write the Lorentz invariants with the help of them.

2. The four-vector potential.

We can do the same kind of calculation on the four vector potential $A_\mu(x_\nu)$ (defined up to a four divergence, a fact which is not difficult to work out, as we will see).The action of the Poincaré group is given by

$$M_{\alpha\beta} = i(x_\alpha \partial_\beta - x_\beta \partial_\alpha) + \Sigma_{\alpha\beta}$$
$$P_\alpha = i\partial_\alpha \tag{E,14}$$

with

$$(\Sigma_{\alpha\beta})_{\sigma\tau} = i \; (g_{\beta\tau} g_{\alpha\sigma} - g_{\beta\sigma} g_{\alpha\tau}) \tag{E,15}$$

The Pauli-Lubanski vector is

$$W^\mu = \frac{1}{2} \varepsilon^{\mu\nu\alpha\beta} \Sigma_{\alpha\beta} \, \partial_\nu \tag{E,16}$$

In order to select a representation of mass zero and helicity s, we must write

$$(W^\mu - s \, P^\mu)_{\sigma\tau} A^\tau = \partial_\sigma \Lambda^\mu \tag{E,17}$$

(the r.h.s. is a four divergence, required by the gauge property. In writing zero, instead, we could not select the expected representation). Eq.(E,17) reads

---

[5]See H. Bacry, Nuov. Cim. *32A*, 448 (1976) for more details.

$$\varepsilon^{\mu\nu\tau\sigma}\partial_\nu A_\tau - i\,s\,\partial^\mu A^\sigma = \partial^\sigma \Lambda^\mu \tag{E,18}$$

This equation implies that the vector $f^\mu = \Lambda^\mu + i\,s\,A^\mu$ obeys

$$\partial^\mu f^\sigma + \partial^\sigma f^\mu = 0 \ \text{ and } \ \partial^\sigma\partial^\sigma f^\mu = 0 \tag{E,19}$$

The last relation (no summation on the index σ) implies that f^μ is of degree one in the x_ν. It follows that its derivatives are constant; finally, we get

$$F_{\mu\nu} + is\,F_{\tilde\mu\nu} = \phi_{\mu\nu} \tag{E,20}$$

where $\phi_{\mu\nu}$ denotes a constant field. In taking the dual of (E,20) and eliminating $F_{\tilde\mu\nu}$, we obtain

$$(1 - s^2)\,F_{\mu\nu} = \phi_{\mu\nu} - is\,\phi_{\tilde\mu\nu} \tag{E,21}$$

Since we are not interested in a constant field, we must take $s^2 = 1$ (as expected) and take the constant field ϕ equal to zero. Eq.(E,20) must be understood as a first order differential equation concerning an eight dimensional object, since s is an operator with two eigenvalues. The eigensubspaces are associated with selfdual and antiselfdual tensors, respectively. This eight component field has $A^0, A^1, A^2, A^3, A^0, -A^1, -A^2, -A^3$ as components.

APPENDIX F

THE ZITTERBEWEGUNG.
THE PRYCE-FOLDY-WOUTHUYSEN TRANSFORMATION.

Let us define the following operator herafter called *the sign of energy*.

$$F = sgn\ H = E^{-1}H = \frac{\alpha.p + m\beta c}{\sqrt{p^2 + m^2 c^2}} \qquad (F,1)$$

$$F^2 = I \qquad (F,2)$$

where I denotes the identity operator and E is the positive operator

$$E = |H| = \sqrt{p^2 c^2 + m^2 c^4} \qquad (F,3)$$

Schrödinger notes that every observable A can be written in a unique way as a sum of an even and an odd part as follows[1]:

$$A = A^+ + A^- \qquad (F,4)$$

$$A^+ = \frac{1}{2}(A + FAF)\ \textit{which commutes with F} \qquad (F,5)$$

$$A^- = \frac{1}{2}(A - FAF)\ \textit{which anticommutes with F} \qquad (F,6)$$

His idea is to apply this decomposition to the usual position operator X; its even part X_S has the property of satisfying the required evolution equation:

$$\frac{dX_S}{dt} = i\ [H,X_S] = H^{-1}p \qquad (F,7)$$

---

[1] The operator F permits to define the projections P_+ (resp. P_-)onto the positive (resp. negative) energy subspaces by the relations:

$$P_{\pm} = \frac{1}{2}(I \pm F)$$

instead of

$$\frac{dX}{dt} = i[H, X] = c\alpha \tag{F,8}$$

which is difficult to accept physically since it implies that the measurement of any speed coordinate leads to one of the two following values: $\pm c$.

Schrödinger considered the operator X_S as describing the mean position of the electron, the difference $X - X_S = \xi$ corresponding to an oscillatory movement (the *zitterbewegung*). The amplitude of this motion is quite small; its order of magnitude is the Compton wavelength, and the period is $\hbar\frac{\pi}{E}$. A simple expression of ξ is given by

$$\xi = -i\frac{\hbar c}{4E}[F, \alpha] \tag{F,9}$$

A simple way to show the properties of X_S and ξ is to use the Dirac evolution operator. If we choose discrete values of time $t_n = n\hbar\frac{\pi}{E}$, we obtain

$$X(t_n) - X(0) = t_n c^2 H^{-1} p \tag{F,10}$$

It is interesting to describe the X decomposition with the aid of the transformation introduced by Pryce [Pry] and which was generalized by Foldy and Wouthuysen [F.W]. It is the unitary transformation which maps the operator F onto β (they have the same spectrum, with the same multiplicities of the eigenvalues). Since in the Dirac representation, β is diagonal, the P.F.W. transformation is, *in this representation,* a way of diagonalizing F, the sign of the energy. Equivalently, it is a way of reducing the representation of the Poincaré group. This unitary transformation is given by

$$U = \sqrt{\frac{2E}{mc^2 + E}}\,\frac{1 + \beta F}{2} \tag{F,11}$$

It is a simple matter to verify that

$$\beta F + F\beta = \frac{2mc^2}{E} \tag{F,12}$$

$$U^2 = \beta F \tag{F,13}$$

and the planned result

$$UFU^+ = \beta \tag{F,14}$$

It is interesting to examine how the generators of the Poincaré group are transformed by the P.F.W. operator. We have

$$UHU^+ = \beta E \tag{F,15}$$

$$UpU^+ = p \tag{F,16}$$

$$UJU^+ = J \quad \text{where } J = X \times p + \frac{\hbar \sigma}{2} \tag{F,17}$$

$$UKU^+ = U(HX + i\frac{\hbar c \gamma_5 \sigma}{2})U^+$$

$$= \beta EX + i\beta\frac{\hbar c^2 p}{2E} - \beta\frac{\hbar c^2 \sigma \times p}{2(E+mc^2)} \tag{F,18}$$

and it is easy to check that all these operators have a diagonal block matrix form in the Dirac representation, which correspond to an "even" form. If we compute UXU^+, we obtain[2]

$$UX_SU^+ = X - \hbar c^2 \frac{\sigma \times p}{2E(E+mc^2)} \tag{F,19}$$

$$U\xi U^+ = -i\hbar c\frac{\beta \alpha}{2E} + i\hbar c^3\frac{\beta(\alpha . p)p}{2E^2(E+mc^2)} \tag{F,20}$$

---

[2]The spin operator s of Chapter 6 cannot be confused with the operator $\frac{1}{2}\hbar \sigma$. The link between the two is given by the relation

$$s = \hbar\frac{\sigma}{2} + (x - q) \times p$$

where q denotes the Newton-Wigner operator for the Dirac particle. Note that x and q are related by the formula $x = UqU^+$. The reader is also invited to compare Eqs (F,19) and (6,8).

Bibliography.

[F.W]   L.L. Foldy and S.A. Wouthuysen, Phys.Rev. *78*, 29 (1950).
[Pry]   M.H.L. Pryce, Proc. Roy. Soc. *A195*, 62 (1948).
[Sch1]   E. Schrödinger, *Collected Papers* (Austrian Academy of Science, Vienna, 1984).

QUOTATIONS.

Page 9, Arago (footnote 3): Two close radiating points, placed on the same vertical line are flashing opposite to a rotating mirror. The rays issued from the upper point are forced to travel through a tube filled with water before reaching the mirror; the rays issued from the lower point reach the reflective surface after a path in a single medium: the air. To be more concrete, we will suppose that the mirror, seen from the observer position, is rotating from right to left. Then, if emission theory is right, if light is matter, the upper point *will seem to be on the left* with respect to the lower point; on the contrary, *it will seem to be on the right* if light results from vibrations of an ethereal medium.

Page 9, Foucault (footnote 4): There is no longer any doubt about the true value of the celerity of light in the vacuum or in our atmosphere. About the speed adopted by light when it penetrates refringent media, it was only given by the calculation which, in interpreting refraction in the emission system or in the wave system, was giving, according to the chosen hypothesis, completely different results. M. Arago, as early as 1938, was the first to recognize the importance of an experiment which, without leading to the exact measure of light celerity in different refringent media, would only set up their difference and, consequently, would permit physicists to know how to interpret refraction.

Page 12, Einstein (footnote 10): A closer reflexion shows us that aether negation is not necessarily required by the special relativity principle. One is allowed to admit the existence of aether, but one has to give up the idea of attributing it a well defined motion, that is one has to strip the aether, by abstraction, of its last mechanical character left by Lorentz. *(translated from French!)*.

Page 51, Schrödinger (footnote 2): The negative eigenvalues [of the energy] have no physical meaning; one would like very much to rid of them. At least, it must be impossible that a "positive" eigenfunction would transform in course of time in giving birth to "negative" functions or at least this variation must be infinitely slow in order to make improbable the enormous change of energy $2mc^2$ we have never observed.

Page 54, Schrödinger (footnote 2): It appears, from the point of view of quantum mechanics, that the theory of relativity lies at the same level as classical mechanics in that it only represents an approximation relative to the macroscopic domain. One will not just admit relativity formulas (for instance, Lorentz formulas) to be valid without change in the intra-atomic domain . They will have to suffer modifications which will be probably analogous to the ones suffered by ordinary mechanics to transform into quantum mechanics. We have to "quantize" the Lorentz transformation.

Page 56, Schrödinger: But the geometrical space is a creation of our imagination, which was transformed by and for the everyday life. It is not sure that Nature possesses this mental construction.

NAME INDEX

Lecture Notes in Mathematics

Vol. 1236: Stochastic Partial Differential Equations and Applications. Proceedings, 1985. Edited by G. Da Prato and L. Tubaro. V, 257 pages. 1987.

Vol. 1237: Rational Approximation and its Applications in Mathematics and Physics. Proceedings, 1985. Edited by J. Gilewicz, M. Pindor and W. Siemaszko. XII, 350 pages. 1987.

Vol. 1250: Stochastic Processes – Mathematics and Physics II. Proceedings 1985. Edited by S. Albeverio, Ph. Blanchard and L. Streit. VI, 359 pages. 1987.

Vol. 1251: Differential Geometric Methods in Mathematical Physics. Proceedings, 1985. Edited by P. L. García and A. Pérez-Rendón. VII, 300 pages. 1987.

Vol. 1255: Differential Geometry and Differential Equations. Proceedings, 1985. Edited by C. Gu, M. Berger and R. L. Bryant. XII, 243 pages. 1987.

Vol. 1256: Pseudo-Differential Operators. Proceedings, 1986. Edited by H. O. Cordes, B. Gramsch and H. Widom. X, 479 pages. 1987.

Vol. 1258: J. Weidmann, Spectral Theory of Ordinary Differential Operators. VI, 303 pages. 1987.

Vol. 1260: N. H. Pavel, Nonlinear Evolution Operators and Semigroups. VI, 285 pages. 1987.

Vol. 1263: V. L. Hansen (Ed.), Differential Geometry. Proceedings, 1985. XI, 288 pages. 1987.

Vol. 1265: W. Van Assche, Asymptotics for Orthogonal Polynomials. VI, 201 pages. 1987.

Vol. 1267: J. Lindenstrauss, V. D. Milman (Eds.), Geometrical Aspects of Functional Analysis. Seminar. VII, 212 pages. 1987.

Vol. 1269: M. Shiota, Nash Manifolds. VI, 223 pages. 1987.

Vol. 1270: C. Carasso, P.-A. Raviart, D. Serre (Eds.), Nonlinear Hyperbolic Problems. Proceedings, 1986. XV, 341 pages. 1987.

Vol. 1272: M.S. Livšic, L.L. Waksman, Commuting Nonselfadjoint Operators in Hilbert Space. III, 115 pages. 1987.

Vol. 1273: G.-M. Greuel, G. Trautmann (Eds.), Singularities, Representation of Algebras, and Vector Bundles. Proceedings, 1985. XIV, 383 pages. 1987.

Vol. 1275: C.A. Berenstein (Ed.), Complex Analysis I. Proceedings, 1985–86. XV, 331 pages. 1987.

Vol. 1276: C.A. Berenstein (Ed.), Complex Analysis II. Proceedings, 1985–86. IX, 320 pages. 1987.

Vol. 1277: C.A. Berenstein (Ed.), Complex Analysis III. Proceedings, 1985–86. X, 350 pages. 1987.

Vol. 1283: S. Mardešić, J. Segal (Eds.), Geometric Topology and Shape Theory. Proceedings, 1986. V, 261 pages. 1987.

Vol. 1285: I.W. Knowles, Y. Saitō (Eds.), Differential Equations and Mathematical Physics. Proceedings, 1986. XVI, 499 pages. 1987.

Vol. 1287: E.B. Saff (Ed.), Approximation Theory, Tampa. Proceedings, 1985–1986. V, 228 pages. 1987.

Vol. 1288: Yu. L. Rodin, Generalized Analytic Functions on Riemann Surfaces. V, 128 pages. 1987.

Vol. 1294: M. Queffélec, Substitution Dynamical Systems – Spectral Analysis. XIII, 240 pages. 1987.

Vol. 1299: S. Watanabe, Yu.V. Prokhorov (Eds.), Probability Theory and Mathematical Statistics. Proceedings, 1986. VIII, 589 pages. 1988.

Vol. 1300: G.B. Seligman, Constructions of Lie Algebras and their Modules. VI, 190 pages. 1988.

Vol. 1302: M. Cwikel, J. Peetre, Y. Sagher, H. Wallin (Eds.), Function Spaces and Applications. Proceedings, 1986. VI, 445 pages. 1988.

Vol. 1303: L. Accardi, W. von Waldenfels (Eds.), Quantum Probability and Applications III. Proceedings, 1987. VI, 373 pages. 1988.

Lecture Notes in Physics

Vol. 287: W. Hillebrandt, R. Kuhfuß, E. Müller, J.W. Truran (Eds.), Nuclear Astrophysics. Proceedings. IX, 347 pages. 1987.

Vol. 288: J. Arbocz, M. Potier-Ferry, J. Singer, V.Tvergaard, Buckling and Post-Buckling. VII, 246 pages. 1987.

Vol. 289: N. Straumann, Klassische Mechanik. XV, 403 Seiten. 1987.

Vol. 290: K.T. Hecht, The Vector Coherent State Method and Its Application to Problems of Higher Symmetries. V, 154 pages. 1987.

Vol. 291: J.L. Linsky, R.E. Stencel (Eds.), Cool Stars, Stellar Systems, and the Sun. Proceedings, 1987. XIII, 537 pages. 1987.

Vol. 292: E.-H. Schröter, M. Schüssler (Eds.), Solar and Stellar Physics. Proceedings, 1987. V, 231 pages. 1987.

Vol. 293: Th. Dorfmüller, R. Pecora (Eds.), Rotational Dynamics of Small and Macromolecules. Proceedings, 1986. V, 249 pages. 1987.

Vol. 294: D. Berényi, G. Hock (Eds.), High-Energy Ion-Atom Collisions. Proceedings, 1987. VIII, 540 pages. 1988.

Vol. 295: P. Schmüser, Feynman-Graphen und Eichtheorien für Experimentalphysiker. VI, 217 Seiten. 1988.

Vol. 296: M. Month, S. Turner (Eds.), Frontiers of Particle Beams. XII, 700 pages. 1988.

Vol. 297: A. Lawrence (Ed.), Comets to Cosmology. X, 415 pages. 1988.

Vol. 298: M. Georgiev, F' Centers in Alkali Halides. XI, 282 pages. 1988.

Vol. 299: J.D. Buckmaster, T. Takeno (Eds.), Mathematical Modeling in Combustion Science. Proceedings, 1987. VI, 168 pages. 1988.

Vol. 300: B.-G. Englert, Semiclassical Theory of Atoms. VII, 401 pages. 1988.

Vol. 301: G. Ferenczi, F. Beleznay (Eds.), New Developments in Semiconductor Physics. Proceedings, 1987. VI, 302 pages. 1988.

Vol. 302: F. Gieres, Geometry of Supersymmetric Gauge Theories. VIII, 189 pages. 1988.

Vol. 303: P. Breitenlohner, D. Maison, K. Sibold (Eds.), Renormalization of Quantum Field Theories with Non-linear Field Transformations. Proceedings, 1987. VI, 239 pages. 1988.

Vol. 304: R. Prud'homme, Fluides hétérogènes et réactifs: écoulements et transferts. VIII, 239 pages. 1988.

Vol. 305: K. Nomoto (Ed.), Atmospheric Diagnostics of Stellar Evolution: Chemical Peculiarity, Mass Loss, and Explosion. Proceedings, 1987. XIV, 468 pages. 1988.

Vol. 306: L. Blitz, F.J. Lockman (Eds.), The Outer Galaxy. Proceedings, 1987. IX, 291 pages. 1988.

Vol. 307: H.R. Miller, P.J. Wiita (Eds.), Active Galactic Nuclei. Proceedings, 1987, XI, 438 pages. 1988.

Vol. 308: H. Bacry, Localizability and Space in Quantum Physics. VII, 81 pages. 1988.

M. Dresden

H. A. Kramers

Between Tradition and Revolution

1987. 16 figures. XXIV, 563 pages. ISBN 3-540-96282-4

Contents: A Remarkable Person in a Very Special Epoch: Why a Biography? The Classical Synthesis: Lorentz. The Unraveling of Classical Physics: Planck and Einstein. Conflicts and the Basic Incompatibility. Bohr: Success and Ambiguity in a Period of Transition. Bohr, the Photon, and Kramers. The Bohr Institute: Pauli and Heisenberg. From Virtual Oscillators to Quantum Mechanics. The Revolution in Progress. – Living Through a Revolution: A Hesitant Start with an Unusual Twist. The Early Copenhagen Years: From Student to Apostle. From Apostle to Prophet: Hints of Trouble to Come. The Multifarious Consequences of a Desperation Revolution. The Curious Copenhagen Interlude. – Waiting for a Revolution that Did Not Happen: The Search for Identity in Changing Times. The Recurrent Theme: Electrons and Radiation. – Kramers as a Person and a Scientist: Conflict or Harmony?: Personality and Style. Obligation and Duty. Kramers' Self Image. Epilogue: Does One Know Better–Understand More? – Appendix: Poems.

K. V. Laurikainen

Beyond the Atom

The Philosophical Thought of Wolfgang Pauli

1988. Approx. 140 pages. Soft cover, in preparation. ISBN 3-540-19456-8

This book is based largely upon a study of the correspondence between Wolfgang Pauli and Markus Fierz. It makes Pauli's philosophy understandable to the average academic reader, not only to the scientist.

From the Foreword:
"It must be emphasized that Pauli's message is revolutionary. He had the distinct opinion that the general trend of Western culture after the 17th century has been one-sided and dangerous. One can characterize it as a vision of a *clockwork* universe – a deterministic world where everything is, in principle, predestined. This vision has created a materialistic culture where the influence of religion has continuously diminished, and of which a strict separation between science and religion is characteristic. Pauli wishes to give us a vision where *spirit* has again been acknowledged as a basic element of the world along with matter. The universe should again be seen as an *organism,* not a clock."

"When the layman says "reality" he usually thinks that he is speaking about something which is self-evidently known; while to me it appears to be specifically the most important and extremely difficult task of our time to work on the elaboration of a new idea of reality. This is also what I mean when I always emphasize that science and religion must have something to do with one another."

Wolfgang Pauli (from a letter to Fierz, 1948 as translated in the book)

Springer-Verlag
Berlin Heidelberg New York
London Paris Tokyo

Springer